NMR

21

Basic Principles and Progress

Editors: P. Diehl E. Fluck H. Günther
R. Kosfeld J. Seelig

Advisory Board: S. Forsén R. K. Harris
C. L. Khetrapal T. E. Lippmaa G. J. Martin
H. Pfeifer A. Pines B. L. Shapiro

V. D. Fedotov   H. Schneider

Structure and Dynamics of Bulk Polymers by NMR-Methods

With 77 Figures

Springer-Verlag Berlin Heidelberg New York
London Paris Tokyo

Professor Dr. Vladimir D. Fedotov
Laboratory of Molecular Biophysics
Institute of Biology, USSR Academy of Science
420084 Kazan, USSR

Professor Dr. Horst Schneider
Technische Hochschule Carl Schorlemmer Leuna-Merseburg, Sektion Physik
Otto-Nuschke-Straße, DDR-4200 Merseburg

ISBN 3-540-50151-7 Springer-Verlag Berlin Heidelberg New York
ISBN 0-387-50151-7 Springer-Verlag New York Berlin Heidelberg

Library of Congress Cataloging-in-Publication Data
Fedotov, V. D. (Vladimir D.), 1940- Structures and dynamics of bulk polymers by NMR-methods / V. D. Fedotov, H. Schneider.
p. cm. - (NMR, basic principles and progress ; 21) Bibliography: p. Includes index.
ISBN 0-387-50151-7 (U.S.)
1. Polymers and polymerization–Analysis. 2. Nuclear magnetic resonance spectroscopy.
I. Schneider, H. (Horst), 1938- II. Title. III. Series. Qd139.P6F43 1989 543'.0877 88-31845 CIP

The publisher cannot assume many legal responsibility for given data, especially as far as directions for the use and the handling of chemicals are concerned. This information can be obtained from the instructions on safe laboratory practice and from the manufacturers of chemicals and laboratory equipment.

Typesetting: Macmillan India Ltd, Bangalore
2151/3140-543210 - Printed on acid-free paper.

Preface

In recent years the development of the NMR method has been closely linked to the creation of the theory of the NMR in solids and with the elaboration on their basic principles and methods of high resolution in solids.

This progress has strongly influenced polymer investigations and has led to the solution of some important questions, growth of the information and to a broadening of the limits of the applicability of this method. From our point of view, the greatest progress in polymer investigations by NMR methods is produced by the solution of the problems concerning (i) the nature of the line-shape, (ii) the roll of the spin–dynamics processes for the longitudinal relaxation and (iii) the measurement of highly resolved NMR spectra of several nuclei and of selective relaxation times. As far as the first two points in particular are concerned, the deuteron resonance plays a permanent growing part in comparison with the proton- and carbon 13-resonances which earlier dominated in polymer investigations. The wide application of elements of the NMR theory in solids for polymer investigations led to a complexity of the experimental interpretation, to an introduction of new conceptions, terms and of a new language. Therefore, the understanding of papers concerning NMR in polymers for non-specialists was rendered more difficult, and the applicability of these often unique results was limited.

The aim of our work is (i) to explain simply and clearly the principles for the understanding of the NMR in solid polymers, (ii) to show what kind of information can be gained using NMR investigations of bulk polymers, and (iii) to demonstrate the ways of obtaining this information, their correctness and their place in comparison with other methods.

We hope that our book will be useful for a large number of readers (physicists, chemists, biologists, engineers etc.) who wish to apply the NMR method or its results to their work in the field of polymers or biopolymers.

The book is made up of three parts. In the first part the more general foundations of the NMR theory and method are discussed (Chapters 1 and 2). In the second the principles and approaches profitable to the analysis of any polymer system are worked out on the basis of a generalization of the results for the simplest systems (Chapters 3 and 4), and lastly some examples of NMR applications for the solution of concrete problems in polymer physics are considered (Chapter 5).

Since the scope of the book is limited, the discussions are restricted to the most important and typical nuclei ^{1}H, ^{13}C, and (briefly) ^{2}D, and, in connection with that, to the dipolar and quadrupolar interaction and to the chemical shift.

Thus such important investigations as the NMR of "other" nuclei (e.g. ^{119}Si, ^{27}Al, ^{31}P, ^{23}Na etc.) in bulk polymers are not included.

Because it is impossible, of course, to present this wide scientific field completely, we refer the reader for more complete information to the secondary literature and to a series of review articles published in recent years and containing numerous references [16, 70, 164, 202].

The present book represents the generalization of the experiences of two scientific groups working in the field of ^{1}H- and ^{13}C-NMR for the investigation of the structure and dynamics in bulk polymers in Kazan (Department of Physics of the Kazan Chemical Technological Institute and Laboratory of Molecular Biophysics of the Biological Institute of the Academy of Science of the USSR) and Merseburg (Department of Physics of the Technical University).

Although there are large distances and boundaries between us leading to some troubles concerning the mutual coordination of our contributions, we were able to overcome these problems. We are specialists in different fields, but the common work of performing and discussing experiments and then writing this book brought us close together. Planned at the beginning as a book consisting of two separated parts, we feel a common responsibility for the book which has developed from this. Whether this was to the advantage for the book we leave this to the readers.

We would like to thank the coworkers in our laboratories for their collaboration over many years making the production of this book possible. These are firstly N. A. Abdrashitova, V. M. Chernov, A. Ebert, G. Griebel, G. Hempel, W.-G. Hiller, G. M. Kadievsky, L. S. Kivaeva, D. Reichert, W. Schenk, K. Schlothauer, G. Simon, A. N. Temnikov, I. M. Vjaselev, M. Wobst. An important role was played by G. Hempel, concerning the investigations of the order and orientation in polymers. Furthermore we thank G. Fleischer, T. N. Khazanovitch and A. I. Maklakov, who contributed to some aspects in Chapter 5, for useful discussions.

Mrs. I. Hoffmann, Miss L. S. Kivaeva, and Mrs. F. Wusterhausen are acknowledged for their technical assistance. Last, but not least, we would like to thank the representatives of the Springer-Verlag for the understanding support during the preparation of the manuscript, and the editor R. Kosfeld for useful advice concerning supplements to the book.

V. D. Fedotov
H. Schneider

Table of Contents

Important Symbols and Abbreviations

General remark:

In this book the international system of units (SI-system) is used. When we mean the magnetic induction denoted with **B** we often speak (although incorrectly) of the magnetic field. Thus this "field" has the SI unit tesla ($1\mathrm{T} = 10^4\,\mathrm{G}$) and some formulas have to be multiplied by a factor $\mu_0/4\pi$, where μ_0 is the magnetic field constant ($\mu_0 = 4\pi \times 10^{-7}\,\mathrm{Vs/Am}$). In order to keep the well-known expressions, this additional factor is written separated at the end of the formulas.

Physical quantities:

$\alpha,\ \beta,\ \gamma$	phase transitions
τ	correlation times
am, cr, int	amorphous, crystalline, intermediate phases
$B_0,\ B_1$	constant and radio-frequency "magnetic field" strength (cf. general remark)
E_a	activation energy
$\mathscr{H}$	Hamiltonian
$\hbar$	Plancks constant divided by 2π
$M_n,\ M_w$	molecular masses (number and weight averaged, always in g mol^{-1})
WLF	Williams–Landel–Ferry relation

NMR quantities:

$\alpha,\ \beta$	parameters in the correlation time distribution
γ	gyromagnetic ratio
δ	anisotropy parameter
$\langle \Delta\omega^2 \rangle$	second moment (always in s^{-2})
η	asymmetry parameter
$\sigma,\ \sigma_{\mathrm{iso}}$	chemical shift and its isotropic mean value
CP	cross-polarization
DD	dipolar-decoupling
D_{s}	self-diffusion coefficient
FID	free induction decay
LF	laboratory frame
LMD	longitudinal magnetization decay
MAS	magic-angle spinning
MW4	cf. Table 2

NOE   nuclear Overhauser effect
RF   rotating frame
rf-pulse   radio-frequency pulse
SE   solid–echo
T_{CH} cross-relaxation time
T_l proton relaxation times (l = 1, 2, 1ϱ, 2e, 1D, 1Z, ZD)
T_l-^{13}C ^{13}C relaxation times
TMD   transverse magnetization decay
WHH4   cf. Table 2

Polymers:

PB   polybutadiene
PBTP   poly(butylene terephthalate)
PC   polycarbonate
PDMS   poly (dimethyl siloxane)
PE, LPE, BPE   polyethylene, linear and branched PE
PEIP   poly(ethylene isophthalate)
PEO   poly(ethylene oxide)
PETP   poly(ethylene terephthalate)
PIB   polyisobutylene
PIP   polyisoprene
PMMA   poly(methyl methacrylate)
PS   polystyrene
PVC   poly(vinyl chloride)
PVF   poly(vinylydene fluoride)
PVME   poly(vinyl methylethylene)

The nuclear magnetic resonance (NMR) phenomenon consists of the resonance absorption of the energy of the radio frequency field by a system of nuclear spins disposed in a constant magnetic field. An interaction of nuclear spins with both the external magnetic field and the fields induced by neighbouring nuclei and molecules takes place during this process. The variety of interactions creates the necessity for high sensitivity of the NMR-method to various characteristics of molecular structure and dynamics.

There are mainly two types of observation: continuous and pulse ones. In the first one, the experiment is carried out on the frequency range and consists of the observation of the *rf* field energy absorption spectra when the *rf* field frequency and the constant magnetic field fulfil the resonance condition. In the second the spin system equilibrium is disturbed by a powerful *rf*-pulse, the frequency spectrum of which covers all the resonance conditions; and the return of the system to the equilibrium is observed on the time scale.

The absorption spectrum and the *rf* pulse-induced time signals, the so-called free induction decays (FID), are connected by a Fourier transformation

$$M(t) = \int_{-\infty}^{+\infty} g(\omega)\exp(-i\omega t)\,d\omega, \tag{1}$$

where $g(\omega)$ is the line shape, and $M(t)$ is the free induction decay.

Recent advances of the pulse methods are connected with the two most important properties of this method. They are the rapid production of the spectrum, and the selective influence on the spin systems. The first increases the sensitivity of the method sharply, the other permits the study of the individual interactions and their time parameters to be carried out separately.

1 Theoretical Aspects

The spin Hamiltonian that determines the main interactions in a solid polymer spin system disposed in the constant magnetic field (B_0) under *rf* field influence (B_1) is

$$\mathcal{H} = \mathcal{H}_Z + \mathcal{H}_{rf} + \mathcal{H}_D + \mathcal{H}_{CS} + \mathcal{H}_Q + \mathcal{H}_J. \tag{2}$$

Here the first and the second terms (the Zeeman- and the *rf*-term) are responsible for the interaction of the nuclear spin system and external magnetic fields $(\vec{B}_0$ and $\vec{B}_1)$. They are determined by the resonance frequency of the effect and by the methods of its observation. The remaining terms are the "internal" ones responsible for interactions that take place within molecular and spin systems. Thus, $\mathcal{H}_{CS}$ describes the spin electronic interaction that determines the line shifts induced by chemical shielding of the nuclei; $\mathcal{H}_D$ and $\mathcal{H}_J$ determines the direct dipole–dipolar nuclear spin interaction and the indirect spin–spin coupling, respectively, and $\mathcal{H}_Q$ – an interaction of nuclear quadrupolar moments with the gradients of the electric fields.

The terms $\mathcal{H}_D$ and $\mathcal{H}_Q$ usually dominate in solids and conceal information on spectral fine structure that is caused by nuclear shielding. In contrast, in low-viscous liquids due to fast isotropic molecular motion, $\mathcal{H}_D$ and $\mathcal{H}_Q$ are averaged and $\mathcal{H}_{CS}$ and $\mathcal{H}_J$ become the dominant terms. All the information is obtained from high-resolution spectra.

For some time, the use of different pulse sequences as well as the mechanical rotation of the sample under the magic angle has made it possible to decrease $\mathcal{H}_D$ sharply and to obtain high-resolution spectra in solids. In this case solid spectra carry additional information in comparison with those of liquids. It occurs due to the anisotropic chemical shift remaining in solid spectra. Every "internal" term of the Hamiltonian can be divided into two parts, secular and nonsecular. The first commutates with the Hamiltonian $\mathcal{H}_Z$ and determines the behaviour of the transverse magnetization component, while the second describes the behaviour of the longitudinal magnetization component and does not commutate with $\mathcal{H}_Z$.

The so-called spin–dynamics effects, caused by processes of energy exchange within the spin system, are usually connected with the relaxation of the transversal magnetization. The isolated spin system develops to the thermodynamic equilibrium under the influence of those processes.

They connect the longitudinal relaxation with the spin–lattice relaxation processes that bring the spin system to the equilibrium with the lattice.

Since in many cases under the influence of different pulse *rf*-fields, the simultaneous excitation of both parts of interaction takes place, the main problem of the experiment is the separation of spin–dynamics and spin–lattice relaxation effects. This problem will be considered in detail in the present book.

1.1   Line Shape and Transverse Relaxation in Solids

1.1.1 Dipole Interaction

The dipolar nuclear spin interaction energy is a function of the absolute value and orientation of the magnetic moment as well as of the length and direction of the vectors that describe the mutual positions of these moments.

The secular part of the dipolar Hamiltonian (expressed in frequency units) for like and unlike spins is given by

$$\mathcal{H}_D^{II} = \sum_{m,n} b_{nm}(\vec{I}_n \vec{I}_m - 3I_{zn}I_{zm}),\tag{3}$$

$$\mathcal{H}_D^{IS} = -2\sum_{m,n} b_{nm}I_{zn}S_{zn},\tag{4}$$

where

$$b_{n,m} = \frac{\gamma_I \gamma_{I(S)}\hbar}{r_{nm}^3}\frac{1}{2}(3\cos^2\theta_{nm} - 1)\frac{\mu_0}{4\pi}.\tag{5}$$

r_{nm} and θ_{nm} mean the magnitude of the internuclear vector connecting spin n with m, and its angle with respect to the magnetic field $\vec{B}_0$, $\vec{I}$ and $\vec{S}$ are the nuclear spin vectors, and I_z and S_z their z-components. γ is the gyromagnetic ratio and $\hbar$ – Planck's constant divided by 2π.

The secular part, mentioned above, may be represented by two terms (see e.g. [1], static (A) and dynamic (B) terms). The first takes into account the interaction of the reference spin with the constant part of the mean local field of the neighbouring spins at the position of this nucleus (more exactly: with the component of this field which is parallel to the external field $\vec{B}_0$). The dynamic term determines the interaction of the reference nucleus with neighbouring nuclei local field components that fluctuate with the Larmor frequency. This resonance interaction provokes the mutual overturn of spins j and k, the so-called flip-flop process, and determines the magnetization flow within the spin system. For interactions of unlike nuclei this term is equal to zero due to its resonance character. The mean local fields created at the sites of the nuclei differ slightly because of their different orientation and different number of neighbours. This fact causes the broadening of the resonance lines. The broadening can be generally compared with the local field that is determined from the second moment $\langle\Delta\omega^2\rangle$ by the formula:

$$B_{\text{loc}}^2 = (3\gamma^2)^{-1}\langle\Delta\omega^2\rangle.\tag{6}$$

The exact calculations of line shape or free induction decay require the solution of the many-body problem, and therefore they present principal difficulties. At present the analytic solutions of this problem exist only in the simplest cases.

They are: two [2] or three [3] spins in arbitrary arrangement; four spins disposed at the corners of a tetrahedron [4] under a fixed orientation with respect to the field $\vec{B}_0$. The exact solutions for the system with a great number of spins were obtained by numerical methods in some special cases. Thus, the papers [5, 6] discussed the linear spin chain, and in [7, 8] the closed chain of five CH_2-groups is considered.

In order to describe the FID or the line shape in solids, simple empiric functions are usually used to approximate the experimental curves. The criterion of approximation correctness is the coincidence of the second and the fourth moments calculated by approximation using Van Vlecks formulae [9] with the moments taken directly from the experimental data. The moments can be calculated from the line-shape function $g(\omega)$:

$$\langle \Delta\omega^{2n} \rangle = \int_{-\infty}^{+\infty} \omega^{2n} g(\omega) d\omega \tag{7}$$

and taking into account the relation (cf. Eq. (1))

$$M(t) = \int_{-\infty}^{+\infty} g(\omega) \cos(\omega t) d\omega \tag{8}$$

from the free induction decay

$$\langle \Delta\omega^{2n} \rangle = (-1)^n \frac{d^{2n}}{dt^{2n}} M(t)|_{t=0}, \tag{9}$$

where each $M(t)$ and $g(\omega)$ are normalized, i.e.

$$M(0) = \int_{-\infty}^{+\infty} g(\omega) d\omega = 1 \tag{10}$$

and $g(\omega)$ is connected with $M(t)$ by

$$g(\omega) = \frac{1}{\pi} \int_0^{\infty} M(t) \cos(\omega t) dt. \tag{11}$$

A series of functions that are most commonly used in the experimental analysis is presented in Table 1.

1.1.2 Chemical Shift

The Hamiltonian of the chemical shifts, which arise because of the different shielding of the applied magnetic field B_0 by the surrounding electrons, has the form

$$\mathcal{H}_{cs} = \gamma \sum_n \vec{I}_n \sigma_n \vec{B}, \tag{12}$$

where $\hat{\sigma}$ is a second rank tensor in the molecular system with the eigenvalues σ_{11}, σ_{22}, σ_{33} which generally describe the chemical shielding anisotropies. When θ and ϕ are the polar angles of the external field $\vec{B}_0$ lying in the z-direction of the laboratory frame (x, y, z) with respect to the principal axis system $(1, 2, 3)$, then the secular part of the chemical shift only observed in high fields is given by

$$\sigma = \sin^2\theta \sin^2\phi \; \sigma_{11} + \sin^2\theta \cos^2\phi \; \sigma_{22} + \cos^2\theta \; \sigma_{33}. \tag{13}$$

Table 1. Typical line shapes in the time and frequency domain and the corresponding second and fourth moments

Denotation	$M(t)\ (t \geq 0)$	$g(\omega)$	$\langle \Delta\omega^2 \rangle$	$\langle \Delta\omega^4 \rangle \cdot$	$\langle \Delta\omega^4 \rangle / \langle \Delta\omega^2 \rangle^2$
Lorentzian	e^{-at}	$\dfrac{1}{\pi}\dfrac{a}{a^2+\omega^2}$			Do not exist
Lorentzian, truncated	$e^{-at}\left(1+\dfrac{2}{\pi}\int_0^t e^{ay}\dfrac{\sin\alpha y}{y}\,dy\right)$ $(\alpha \gg a)^*$	$\dfrac{1}{\pi}\dfrac{a}{a^2+\omega^2}\ \text{ if } |\omega| \leq \alpha$ $0 \qquad\qquad \text{ if } |\omega| > \alpha$	$\dfrac{2a\alpha}{\pi}$	$\dfrac{2a\alpha^3}{3\pi}$	$\dfrac{\pi}{6}\left[\dfrac{\alpha}{a}+\pi-\dfrac{2}{\pi}+\dfrac{a}{\alpha}\left(\dfrac{3}{4}\pi^2-7\right)\right]^{**}$
Gaussian	$e^{-\frac{1}{2}(at)^2}$	$\dfrac{1}{\sqrt{2\pi}\,a}\exp\left[-\dfrac{1}{2}\left(\dfrac{\omega}{a}\right)^2\right]$	a^2	$3a^4$	3
Pakes doublett	$e^{-\frac{1}{2}(at)^2}\cos(bt)$	$\dfrac{1}{2\sqrt{2\pi}\,a}\left\{\exp\left[-\dfrac{1}{2}\left(\dfrac{\omega+b}{a}\right)^2\right]+\exp\left[-\dfrac{1}{2}\left(\dfrac{\omega-b}{a}\right)^2\right]\right\}$	a^2+b^2	$3a^4+6a^2b^2+b^4$	$3-2\left[\left(\dfrac{a}{b}\right)^2+1\right]^{-2}$
Abragam	$e^{-\frac{1}{2}(at)^2}\dfrac{\sin(bt)}{bt}$	$\dfrac{1}{2b}\left[\mathrm{erf}\left(\dfrac{b+\omega}{\sqrt{2}\,a}\right)+\mathrm{erf}\left(\dfrac{b-\omega}{\sqrt{2}\,a}\right)\right]^{***}$	$a^2+b^2/3$	$3a^4+2a^2b^2+b^4/5$	$3-\dfrac{2}{15}\left[\left(\dfrac{a}{b}\right)^2+\dfrac{1}{3}\right]^{-2}$

* If $\alpha \gg a$ does not hold true, π has to be replaced by $2\arctan\dfrac{\alpha}{a}$ in the expressions for $M(t)$ and $g(\omega)$, and the expressions for $\langle \Delta\omega^2 \rangle$ and $\langle \Delta\omega^4 \rangle$ are

$a\alpha\left[\arctan\dfrac{\alpha}{a}\right]^{-1}-a^2$ and $a\alpha^3\left[3\arctan\dfrac{\alpha}{a}\right]^{-1}\left[1-3\left(\dfrac{a}{\alpha}\right)\right]^2+a^4$ \quad ** For $\alpha \geq 2a$ the relative error is smaller than 2.5×10^{-4} \quad *** $\mathrm{erf}\,x=\dfrac{2}{\sqrt{\pi}}\int_0^x e^{-t^2}\,dt$

Using the parameters for the isotropic chemical shift

$$\sigma_{\mathrm{iso}} = \tfrac{1}{3}(\sigma_{11} + \sigma_{22} + \sigma_{33}), \tag{14}$$

for the anisotropy

$$\delta = (\sigma_{33} - \sigma_{\mathrm{iso}}), \tag{15}$$

and for the asymmetry

$$\eta = \frac{1}{\delta}(\sigma_{22} - \sigma_{11}) \tag{16}$$

we conclude

$$\sigma = \sigma_{\mathrm{iso}} + \frac{\delta}{2}[(3\cos^2\theta - 1) + \eta \sin^2\theta \cos 2\phi]. \tag{17}$$

The line shape $I(\sigma)$ in a crystalline sample can be derived using (13) or (17), respectively, where the effective chemical shift must be weighted according to the isotropic probability distribution, as is shown by Bloembergen and Rowland [10] and in another form by Haeberlen [11]. Under the assumption

$$\sigma_{11} \leq \sigma_{22} \leq \sigma_{33} \quad (\text{i.e. } 0 \leq \eta \leq 3) \tag{18}$$

follows

$$I(\sigma) = a^{-1} K\left(\frac{b}{a}\right) \quad \text{for} \quad \sigma_{11} \leq \sigma \leq \sigma_{22}, \tag{19a}$$

$$I(\sigma) = b^{-1} K\left(\frac{a}{b}\right) \quad \text{for} \quad \sigma_{22} \leq \sigma \leq \sigma_{33}, \tag{19b}$$

and

$$I(\sigma) = 0 \quad \text{otherwise}, \tag{19c}$$

where

$$a = \pi[(\sigma_{22} - \sigma_{11})(\sigma_{33} - \sigma)]^{1/2}, \tag{20a}$$

$$b = \pi[(\sigma_{33} - \sigma_{22})(\sigma - \sigma_{11})]^{1/2}, \tag{20b}$$

and $K(k)$ is the complete elliptical integral of the first type

$$K(k) = \int_0^{\pi/2} \frac{d\gamma}{(1 - k^2 \sin^2\gamma)^{1/2}}. \tag{21}$$

For this adsorption line shape, shown in Fig. 1a, the intensity ratio at the ends is given by the inverse ratio of the roots of the distances from the singularity at σ_{22}

$$\frac{I(\sigma_{11})}{I(\sigma_{33})} = \left[\frac{\sigma_{33} - \sigma_{22}}{\sigma_{22} - \sigma_{11}}\right]^{1/2} = \left[\frac{3 - \eta}{2\eta}\right]^{1/2}. \tag{22}$$

The difference of these distances normalized by the total anisotropy yields the difference between the normalized areas to the left and to the right of the singularity (cf. Fig. 1(a)) according to

$$\sin\left[\frac{\pi}{2}\frac{A_2 - A_1}{A_1 + A_2}\right] = \frac{(\sigma_{33} - \sigma_{22}) - (\sigma_{22} - \sigma_{11})}{(\sigma_{33} - \sigma_{22}) + (\sigma_{22} - \sigma_{11})} = \frac{3(1 - \eta)}{3 + \eta}, \tag{23}$$

which was derived using Eq. (3–15) in [11].

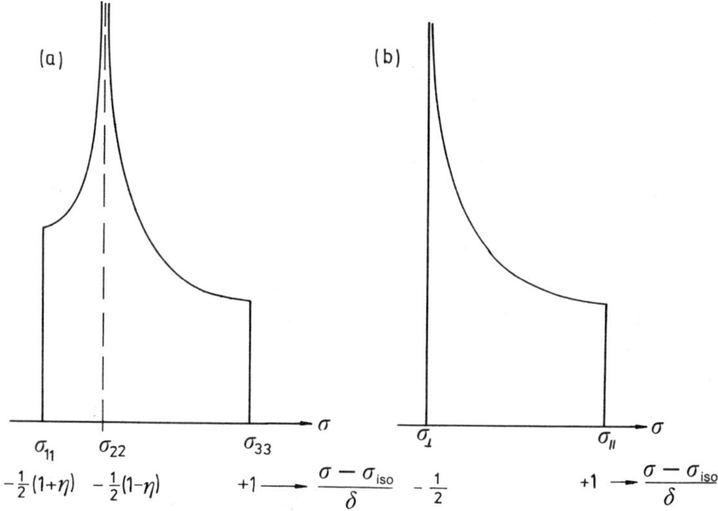

Fig. 1. Line shapes for polycrystalline samples with (**a**) a general ($\eta = \frac{1}{2}$) and (**b**) axially symmetric ($\eta = 0$) chemical shielding tensor σ, given by Eqs. (19–21) and (28)

For the first moments of the line shapes with respect to the isotropic value

$$\langle \Delta\omega^n \rangle = \int_{\sigma_{11}}^{\sigma_{33}} I(\sigma)(\sigma - \sigma_{\mathrm{iso}})^n \, d\sigma \tag{24}$$

it follows for $n = 0, 1, 2, 3$

$$\langle \Delta\omega^n \rangle = 1, 0, \frac{\delta^2}{15}(3 + \eta^2), \ \frac{2\delta^2}{35}(1 - \eta^2), \tag{25}$$

respectively, providing the possibility to determine δ and η from the second and the third moments. This is of particular interest, because (i) these lower moments do not depend on the rotational frequency by magic angle spinning [12] and can be measured by special techniques (cf. Sect. 3), and (ii) the evaluation of the moments from the experimental curves does not require knowledge of the isotropic mean value σ_{iso}. Denoting the moments of the normalized experimental line with respect to an arbitrary value of σ by M_n one can easily derive the following connections

$$\langle \Delta\omega^2 \rangle = M_2 - M_1^2, \tag{26a}$$

$$\langle \Delta\omega^3 \rangle = M_3 - 3M_2 M_1 + 2M_1^3 \tag{26b}$$

and hence, the parameters δ and η carrying the whole anisotropy information were obtained in this way. For a tensor with uniaxial symmetry with

$$\sigma_{11} = \sigma_{22} = \sigma_\perp, \quad \sigma_{33} = \sigma_\| \quad \text{(i.e. } \eta = 0) \tag{27a}$$

and also for

$$\sigma_{11} = \sigma_\|, \quad \sigma_{22} = \sigma_{33} = \sigma_\perp \quad \text{(i.e. } \eta = 3) \tag{27b}$$

follows the more simple expression for the line shape

$$I(\sigma) = \tfrac{1}{2}[(\sigma_{||} - \sigma_\perp)(\sigma - \sigma_\perp)]^{-1/2}, \tag{28}$$

which is depicted in Fig. 1b.

For both types of symmetry the shielding tensor elements can be read directly from the line shape. A very interesting case is the change of the shielding tensor for a molecule subjected to a very anisotropic molecular motion, e.g. to a rapid restricted rotation about one axis in the molecule. The partial averaging process produces the axial asymmetry. If this rotation takes place about a principal axis, which often happens, then $\sigma_{||}$ of the now axially symmetric tensor lies in that direction and has the value of the corresponding former principal element, whereas $\sigma_\perp$ results as the mean value of the other two components.

Even for polycrystalline samples these theoretical powder patterns have been observed very seldom. For the usually semicrystalline or amorphous polymers there are many reasons for a distribution of the shielding tensor elements leading [13–15] to a smoothing of the curves shown in Fig. 1. These distributions $k_i(\sigma_{ii})$ of shielding tensor elements σ_{ii} with the lower and upper limit σ_{ii}^l and σ_{ii}^u yield a measured line shape $G(\sigma)$ given by

$$G(\sigma) = \int\limits_{\sigma_{11}^l}^{\sigma_{11}^u} d\sigma_{11} \int\limits_{\sigma_{22}^l}^{\sigma_{22}^u} d\sigma_{22} \int\limits_{\sigma_{33}^l}^{\sigma_{33}^u} d\sigma_{33}\, k_1(\sigma_{11})k_2(\sigma_{22})k_3(\sigma_{33})I(\sigma, \sigma_{11}, \sigma_{22}, \sigma_{33}). \tag{29}$$

We get another change of the line shape for oriented polymer samples, for which e.g. in the axially symmetric case the distribution function $f(\cos\theta)$ is not equally distributed as it is realized in the isotropic case. Then the line shape $G(\sigma)$ follows according to

$$|G(\sigma)d\sigma| = |f(\cos\theta)d(\cos\theta)| \tag{30}$$

and for an equal distribution, i.e. $f(\cos\theta) = 1$,

$$I(\sigma) = \left|\frac{d(\cos\theta)}{d\sigma}\right| \tag{31}$$

which yields with

$$\sigma = \sin^2\theta\,\sigma_\perp + \cos^2\theta\,\sigma_{||}, \tag{32}$$

following from Eqs. (13) and (27), the theoretical powder spectrum $I(\sigma)$ in the isotropic case given in Eq. (28). Therefore, for anisotropic materials with a distribution of $(\cos\theta)$ the line shape is given by [13, 15]

$$G(\sigma) = f(\cos\theta)I(\sigma). \tag{33}$$

The determination of $f(\cos\theta)$ is also possible using point by point the division of $G(\sigma)$ and $I(\sigma)$ taking into account the relation following from Eq. (32)

$$\cos\theta = \left[\frac{\sigma - \sigma_\perp}{\sigma_{||} - \sigma_\perp}\right]^{1/2}. \tag{34}$$

1.1.3 Quadrupolar Interaction

The Hamiltonian for the quadrupolar interaction which arises because of the quadrupolar coupling of nuclear species (having $I > \frac{1}{2}$ spin) to the electric field gradient, has the form (here written for one nucleus) [1]

$$\mathcal{H}_Q = \frac{e^2\,qQ}{4\hbar I(2I-1)}\,[3I_z^2 - I(I+1) + (I_x^2 - I_y^2)],\tag{35}$$

where e is the elementary charge and eQ the electric quadrupole moment. With the tensor components of the second derivatives of the electrostatic potential V_{xx}, V_{yy}, V_{zz} (with $|V_{zz}| \geq |V_{xx}| \geq |V_{yy}|$), the electric field gradient q and the asymmetry parameter η, respectively, are given by

$$q = \frac{1}{e}\,V_{zz}\tag{36}$$

and

$$\eta = \frac{V_{yy} - V_{xx}}{V_{zz}}\quad(\text{i.e.:}\ 0 \leq \eta \leq 1).\tag{37}$$

Using first order perturbation theory in the case which is of interest here $\mathcal{H}_Q \ll \mathcal{H}_z$, for the frequencies of the two allowed NMR transitions for $I = 1$ spin nuclei [16, 17], follows

$$\omega = \omega_0 \pm \delta\,(3\cos^2\theta - 1 - \eta\sin^2\theta\cos 2\phi),\tag{38}$$

where θ and ϕ are the polar angles of the external field with respect to the principal axes system of the electrical field gradient tensor and

$$\delta = \frac{3}{8}\frac{e^2\,qQ}{\hbar}.\tag{39}$$

The latter quotient (without the numerical factor) is the quadrupole coupling constant. A centre line at the frequency ω_0 occurs only for nonintegral spin quadrupolar nuclei which we shall not take into consideration in this book.

In a similar manner as for chemical shifts we can get the line shape for an isotropic polycrystalline sample, only the sign of η is changed and two symmetric components are added. Figure 2 shows these theoretical line shapes for $\eta = 0, \frac{1}{2}$, and 1, respectively, which correspond to Fig. 1 (see also the significant values for $(\sigma - \sigma_{\text{iso}})/\delta$ and Eqs. (17, 22)).

In a similar manner as for the line shape due to the chemical shift anisotropy, this powder pattern is smoothed for many reasons and strongly changed for oriented samples in which the principal axes of the electric field gradient tensor with respect to the external magnetic field are not equally distributed.

Another change of the line shape, given here in absence of motion only, will occur under the influence of molecular motion depending on the type as well as on the timescale of this process. This will be discussed in Sect. 1.3.5.

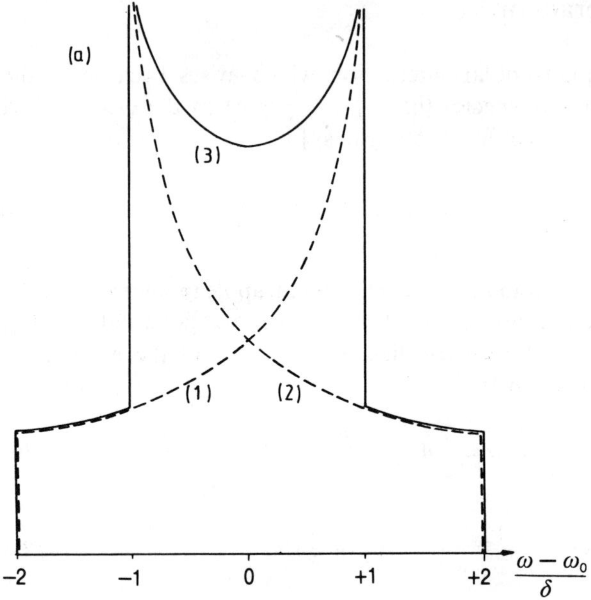

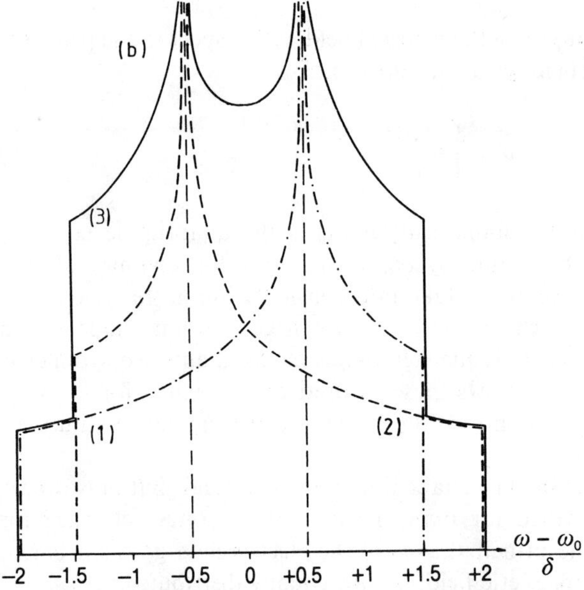

Fig. 2. Line shapes for polycrystalline samples with first order quadrupole coupling with (**a**) an axially symmetric ($\eta = 0$) and (**b**) a general ($\eta = \frac{1}{2}$) electrical field gradient tensor and (**c**) for the limiting case ($\eta = 1$). *1*, *2*: two symmetric components, *3*: their sum

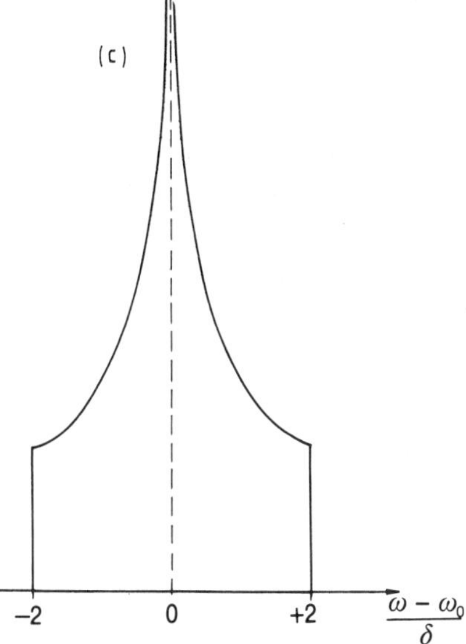

Fig. 2c

1.1.4 Line Narrowing in Solids

To perform high resolution NMR experiments it is necessary to change the too complicated Hamiltonian (cf. e.g. Eq. (2)) in such a way that the unimportant part for a special problem is suppressed and the part which we are interested in influences the result of the measurement. Elimination proceeds making the unwanted part of the Hamiltonian time-dependent, in this way selectively averaging that part [11]. There are two very important natural averaging processes (i) via the thermal motion in liquids, which leads to the vanishing of the dipolar Hamiltonian and to the possibility of measuring chemical shifts and indirect nuclear spin–spin couplings and (ii) via the rapid variation of the nonsecular terms of the Hamiltonian in high magnetic fields B_0 in a rigid lattice which causes the truncation of the Hamiltonian.

The possibilities for the line narrowing in solids can be derived e.g. from Eqs. (3–5) for the dipolar Hamiltonian. It means the dipolar, the chemical shift anisotropy and possibly other Hamiltonians can be averaged. That causes linewidths of several kilohertz to a few hertz, necessary for the investigation of chemical shifts or spin–spin couplings.

(i) Magic–Angle Sample–Spinning Experiment

Under the influence of a fast sample–spinning the spherical harmonics of the second order are modulated by a factor $\frac{1}{2}(3\cos^2\beta - 1)$, where β is the angle of the rotation axis with respect to the external magnetic field, yielding the vanishing of this term for the magic angle $\beta_m = 54°\,44'$ [18, 19].

For the chemical shift e.g. one gets from Eq. (17) neglecting all terms which change with the high sample rotation frequency ω_r, since they do not contribute to the spectrum

$$\sigma = \sigma_{\text{iso}} + \frac{\delta}{4}(3\cos^2\beta - 1)[(3\cos^2\theta' - 1) + \eta\sin^2\theta'\cos 2\phi'], \qquad (40)$$

where θ' and ϕ' are the spherical polar angles of the specimen rotation axis with respect to the principal axis of the shift tensor. This means the line shape in polycrystalline materials is not influenced by sample rotation; however, the line width, i.e. the anisotropy, is reduced by the factor $\frac{1}{2}(3\cos^2\beta - 1)$ mentioned above. It can be shown [11] that all other anisotropic interactions will also be removed, if the two following assumptions are fulfilled: restriction to secular terms is possible (strong magnetic fields and only slow molecular motion) and the spinning speed is high enough compared with the considered interaction. The rotational speed necessary for the elimination of the dipolar interaction ($\gtrsim 10\,\text{kHz}$) is difficult to realize. However, if the removal of this interaction is possible by means of other methods (c.f. (ii) and (iii)) the MAS technique averages the only remaining anisotropy of the chemical shift using spinning frequencies of 2 to 3 kHz and allows the measurement of its isotropic mean value, which was demonstrated first by Schaefer and coworkers [20]. The possibility of the nearly complete removal of the deuterium quadrupole coupling and therefore, for the observation of high-resolution deuterium solid-state spectra using the MAS-technique was also shown by Pines and others [21].

(ii) Multiple-Pulse NMR

Several multiple-pulse cycles produce a modulation of the spin part $(\vec{I}_1\vec{I}_2 - 3I_{z1}I_{z2})$ of the dipolar Hamiltonian and assuming an appropriate choice of widths, phases and distances of the rf-pulses lead to a time averaging in the spin space. The four-pulse-cycle WHH4 given in the pioneering work of Waugh, Huber and Haeberlen [22] represents the basic principles of time averaging and has often been modified for better resolution under real experimental conditions (cf. e.g. [11, 23]).

For the elimination of the heteronuclear dipolar interaction additional precautions must be taken. The anisotropy of chemical shift is poorly reduced by a scaling factor and has to be removed using the MAS technique. This method is mainly applied for protons; it is also used for ^{19}F- and ^{31}P-measurements.

(iii) Rare Spins and Dipolar Decoupling of the Abundant Nuclei

Finally we use the first part of the dipolar Hamiltonian in Eq. (5). The strong r^{-3}-dependence shows the weak contribution of this term in the investigation of rare spins under natural abundance as e.g. ^{13}C (1.1%). However, often non-observed abundant spins (e.g. protons) yield strong heteronuclear dipolar interaction (cf. Eq. (4)). A strong rf-field at the resonance frequency of the abundant spins leads to their saturation and the time average of the heteronuclear coupling vanishes, if the rf-field B_{1I} is so strong that $\omega_{1I} = \gamma_I B_{1I}$ is large compared with the strength of the heteronuclear coupling, i.e. this rf-field has to be much stronger than that for the decoupling of indirect spin–spin interaction in liquids. This decoupling field can be used simultaneously for the cross-polarization of the rare spins via the contact with the abundant spins. This method proposed by Waugh and coworkers [24] and called proton-enhanced nuclear induction spectroscopy is described in Sect. 1.2.6.1. As in the case of the multiple-pulse technique additional use of MAS eliminates the anisotropy of chemical shifts and leads to highly resolved spectra.

1.2 Spin–Dynamics and Longitudinal Relaxation

1.2.1 Spin Temperature

Since spin–spin relaxation times (T_2) are much shorter than those of spin–lattice relaxation (T_1) in solids, the spin system can be considered isolated from the lattice. Due to this fact, a spin temperature different from the lattice temperature may be ascribed to this system. Naturally, the spin temperature concept makes sense only when the equilibrium within a spin system is reached, i.e. when the transverse magnetization vanishes.

According to the NMR theory [25, 26] the solid spin system placed into a magnetic field may be described by two reservoirs with the thermal heat capacities proportional to the square of the magnetic field. One of the reservoirs, called Zeeman, corresponds to the energy of spin interaction with the external magnetic field, the other, called dipolar, corresponds to the secular part of dipolar Hamiltonian $\mathcal{H}_D$.

Under certain conditions reservoirs may have different spin temperatures and therefore their own spin–lattice relaxation time $(T_1$ and $T_{1D})$. Under other conditions caused by nonsecular terms the intensive energy transfer between these two subsystems, the so-called Zeeman–dipolar cross relaxation, can take place resulting in a unified spin temperature. The spin system united in such a way relaxes to a lattice with the weight-averaged rate of spin–lattice relaxation. The thermal heat capacities of each subsystem are the weighting factors in this case. As can be seen in Sect. 3.2, for $T_{1\varrho}$ measurements the first case is usually observed at $B_1 \gg B_{\text{loc}}$ while the second one is observed at $B_1 \approx B_{\text{loc}}$. The typical solid spin system diagram is presented in Fig. 3. The path of spin–spin and spin–lattice relaxation is shown by arrows.

The spin system of samples with different magnetic nuclei includes Zeeman and dipolar reservoirs according to the number of different nuclei. In general, the exact establishment of the spin system structure is an important problem of the experimental analysis.

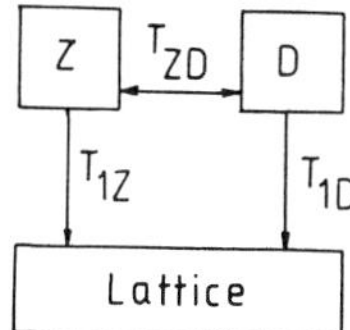

Fig. 3. Structure of spin-systems in solids for spin $\frac{1}{2}$ nuclei

1.2.2 Spin Diffusion

The spin diffusion concept is associated with the process of nuclear magnetization transfer within the spin system that takes place according to the dynamic term of the dipole–dipole interaction (cf. e.g. [27, 28]). In [29] it is shown that spin and material diffusion processes in spite of their quite different nature can be nominally described by the same functions (diffusion equations). These equations include the diffusion coefficient, that is the sum of spin (D_s) and material (D_m) diffusion coefficients here. D_s is connected with the mean square transfer distance (as is D_m) expressed by

$$\langle x^2 \rangle = 6D_s t, \tag{41}$$

where t is the observation time that is determined in NMR-experiments by the spin–lattice relaxation time or by the pulse distance. It is shown in [28], that

$$D_s = a \frac{r^2}{T_2}, \tag{42}$$

where r is the distance of the neighbouring spin, T_2 is the spin–spin relaxation time, a is a numeric factor depending on the crystal lattice type (it is equal to 0.1 for the cubic lattice). From Eq. (42) it follows that D_s decreases with the rise of temperature or remains constant contrary to D_m, which, as is known increases. Spin diffusion has the greatest influence on spin–lattice relaxation in a heterogeneous system. This influence results in an excess of spin energy transition to nuclei with strong coupling with the lattice. Paramagnetic impurities, lattice defects, side and end methyl groups rotating rapidly, etc. may be such hot points. Besides, an effective energy transfer between spin systems in different structural polymer formations including polymer chains of different mobility and therefore having different lattice coupling may take place due to the spin diffusion mechanism. Here magnetization transfer is mainly limited by the transition over the boundary. In contrast to the typical spin diffusion that takes place in macroscopically homogeneous samples this process can be named interphase spin or material diffusion. In order to describe the spin diffusion influence on the nuclear relaxation quantitatively, one-dimensional models of the diffusion to the defects and of diffusion of the defects themselves for microscopically heterogeneous systems have been used [30, 31, 332]. Moreover, macroscopically heterogeneous systems have been described by the kinetic equations for magnetization transfer between discrete spin systems that are analogous to the well-known McConnel equations for chemical exchange [154, 155, 212].

If the diffusion rate is small compared with the spin–lattice relaxation rates, then spin–diffusion has no influence on the measured relaxation times. In this case the signals related to individual species (atomic groups, separate molecules or different structure phases) can be measured selectively. In the opposite case only an averaged signal of the whole sample can be observed. According to Eq. (41), the shorter the observation time determined by magnetic relaxation time in many experiments, the shorter is the spin diffusion spreading distance and the less effective is the averaging of relaxation characteristics. Due to this fact magnetic relaxation times in relation to their sensitivity can be summarized in the sequence

$$T_2 < T_{1D} < T_{1\varrho} < T_1. \tag{43}$$

1.2.3 Spin–Flip Effects

The measurement of the ^{13}C spin–lattice relaxation time $T_{1\varrho}$ in the rotating frame is not only influenced by the contact with the lattice which we are interested in but also by the contact with the dipolar system of the protons. The latter effect is caused by fluctuations of proton spin pairs without energy exchange, i.e. with a high frequency of about 10 kHz. These spin flips lead to modulations of the local field at the carbon sites and yield additional transitions between the ^{13}C Zeeman levels in the rotating frame of coordinates [32]. The measured relaxation rate $T_{1\varrho}^{-1}$ is given by

$$T_{1\varrho}^{-1} = (T_{1\varrho}^L)^{-1} + (T_{CH}^0)^{-1}, \tag{44}$$

where $T_{1\varrho}^L$ characterizes the spin–lattice interaction and T_{CH}^0 the spin–spin process described above. To decide the relevance of the observed $T_{1\varrho}$ for the description of the molecular motion it is necessary to estimate the strength of the T_{CH}^0 value (which is equivalent to the cross-polarization time in the case of proton adiabatic demagnetization in the rotating frame and similar to the cross-relaxation time T_{CH} explained in Sect. 2.1) using an equation given by Cheung and Yaris [33]:

$$(T_{CH}^0)^{-1} \approx 51.2\, v_{loc} \exp\left(-\sqrt{3}\,\frac{v_{1C}}{v_{loc}}\right) \cdot \overline{\tfrac{1}{2}[3\cos^2\alpha(t) - 1]}^{\,t}, \tag{45}$$

where v_{loc} is the local field (in frequency units) at the carbon site, $v_{1C} = \frac{1}{2\pi}\gamma_C B_{1C}$ and the time average has to be formed with respect to $\alpha(t)$, which is the angle between the dipolar vectors at the times $t = 0$ and t, respectively (e.g. $\overline{\tfrac{1}{2}[3\cos^2\alpha(t) - 1]}^{\,t} = 1$ for a rigid lattice). A very strong decrease of T_{CH}^0 with the strength of the local field and a drastic increase with the increase of the carbon *rf*-field strength and under the influence of the molecular motion are typical. This is caused by the influence of the local field on the time average term.

1.2.4 Spin-Locking Experiment

Spin-locking experiments in solids consisting of equal spins ($I = 1/2$) are described in this section. The structure of the spin system in those solids is presented in Fig. 3. The 90° pulse turns the magnetization M_0 (being initially parallel to $\vec{B}_0$) to the x-axis of the frame rotating with the frequency $\omega_0 = \gamma B_0$. The following pulse at the resonance field $\vec{B}_1$ is applied just along that x-axis to preserve M_0 parallel to $\vec{B}_1$. Thereby the Zeeman reservoir energy decreases $(B_0/B_1)^2$-times, and the equilibrium within the spin system is disturbed. The spin system equilibrium is established in two steps. In the first the direct energy exchange between Z- and D-reservoirs displayed by rapid damping of magnetization oscillations takes the system to quasi-equilibrium with the establishment of spin temperatures (α_0^{-1} and β_0^{-1}) in both reservoirs. In the second step a common temperature with the corresponding final magnetization M_f (cf. e.g. [25, 34]) is established by the Zeeman-dipolar cross-relaxation process

$$M_f = M_0 \frac{B_1^2}{B_1^2 + B_{loc}^2}. \tag{46}$$

Then the spin–lattice relaxation process takes the system to the equilibrium with the lattice.

The most complete description of the first step is presented in [34], while the second and the third ones are described in [35]. The case of spin–lattice and Zeeman dipolar relaxation times being close to each other, is of great importance for polymer analysis [36]. That work presents solutions of the spin temperature kinetic equations describing the establishment of the equilibrium under spin-locking conditions:

$$\frac{d\alpha}{dt} = \frac{B_1^2}{B_{loc}^2} T_{ZD}^{-1}(\alpha - \beta) + T_{1D}^{-1}\alpha, \tag{47a}$$

$$\frac{d\beta}{dt} = T_{ZD}^{-1}(\beta - \alpha) + T_{1Z}^{-1}\beta. \tag{47b}$$

α and β are the reciprocal spin temperatures of dipolar and Zeeman reservoirs. The relaxation times T_{1D}, T_{1Z}, and T_{ZD} are explained in Fig. 3.

The solution of Eqs. (47) for β can be written as:

$$\beta = M(t) = A_1 \exp\left(-\frac{t}{T_{1\varrho1}}\right) + A_2 \exp\left(-\frac{t}{T_{1\varrho2}}\right), \tag{48}$$

where the quantities $T_{1\varrho1,2}$ and $A_{1,2}$ are functions of the spin system parameters T_{ZD}, T_{1Z}, T_{1D}, B_1, B_{loc}, α_0 and β_0. (The functions are presented in [36]).

Hence, one can see that the magnetization decay even in a homogeneous solid sample is a sum of three components under certain conditions ($B_1 \approx B_{loc}$). Two of the components, the oscillating and the short one, are generally determined by the spin dynamic process, while the slow decay relates only to the spin–lattice relaxation. If this fact is omitted, the interpretation of the experiment for multiphase systems may be incorrect.

When the condition $B_1 \gg B_{loc}$ is fulfilled, the difficulties of interpretation disappear, but as it has been shown [37, 38], the experiments in small B_1 are of great importance in the analysis of heterogeneous systems. It is connected with the fact that in low fields the most rapid spin–lattice relaxation takes place and the influence of interphase spin diffusion is weaker, thus the true relaxation times are observed.

The analysis showed that the long time $T_{1\varrho1}$ is mainly determined by spin–lattice relaxation times (T_{1Z} and T_{1D}). $T_{1\varrho1}$ tends to T_{1Z} for the limit $B_1 \gg B_{loc}$ and to T_{1D} for the limit $B_1 \ll B_{loc}$. The short time $T_{1\varrho2}$ approaches T_{ZD} for $B_1 \ll B_{loc}$ and reduces to T_{1D} at the opposite limit. On the other hand, in the last case the intensity of the short component reduces to zero and the magnetization decays exponentially with the relaxation time $T_{1\varrho 1} = T_{1z}$, as it was shown in the experiment.

1.2.5 Multipulse Spin-Locking

Recently, multipulse sequences with an effect on the spin system similar to that of the locking pulse were used to study molecular motions in polymers [39, 40]. The theory describing the action of these sequences on the spin system with mobile nuclei has been worked out in [41]. There it is demonstrated that in systems with rapid isotropic rotation of spins, the magnetization decay time for these sequences (T_{2e}) and the spin–lattice relaxation time $T_{1\varrho}$ coincide to an accuracy of a factor $\pi^2/8$. A similar agreement also yields for more complicated correlation functions [42].

The rigorous solution of the effect of multipulse sequences on immobile spin solids is given in [43, 44] while the effects on the mobile spin solids are described in [45–48]. It has been demonstrated that the agreement between T_{2e} and $T_{1\varrho}$ is not given in each case. So there is always an additional magnetization decay caused by spin dynamic processes in the case of slow isotropic rotation and/or local aniso-tropic movements, especially if the condition $\tau_c/2\tau \ll 1$ is not fulfilled (τ_c is the correlation time of the motion and τ a characteristic pulse distance). This damping may dominate in some cases. Hence, the application of these multipulse sequences to spin–lattice relaxation time measurements in the rotating frame always requires an additional analysis and a strict fulfilment of various limiting conditions.

1.2.6 Cross-Polarization and Nuclear Overhauser Effect

1.2.6.1 Cross-Polarization

Here we will briefly discuss one of the many possible double resonance experiments [24] using the cross-polarization of rare (e.g. ^{13}C) and abundant (e.g. ^{1}H) spins (i.e. for the relation between the spin numbers $N_H \gg N_C$). The coupling of these two spin systems to the lattice is characterized by the spin–lattice relaxation times T_1^H and T_1^C, respectively, and the coupling of both spin systems together by the dipolar cross-relaxation time T_{CH}. The four steps of the CP-experiments are:

 (i) An approach of the proton magnetization to its thermal equilibrium value.
 (ii) Storing of the proton magnetization in the rotating frame using a spin-locking experiment.
(iii) Cross-relaxation between H- and C-spins by irradiation of the two frequencies under the Hartmann–Hahn [49] condition

$$\gamma_C B_{1C} = \gamma_H B_{1H} \quad (\text{i.e. } \omega_{1C} = \omega_{1H}), \tag{49}$$

where B_{1C} and B_{1H} are the amplitudes of the *rf*-fields of the C- and H-spins, respectively. Consequently, a rapid polarization transfer under energy con-servation (in the rotating frame) is possible.

(iv) Observation of the carbon free induction decay without the B_{1C}-field but under proton decoupling by retaining irradiation of the B_{1H}-field.

The theoretical value for the increase of magnetization by this technique compared with the usual free induction experiment with proton spin decoupling is about γ_H/γ_C (experimentally one often reaches factors of about 2 . . . 3 instead of nearly 4). The time of carbon cross-polarization is characterized by the cross-relaxation time T_{CH} and is limited by the (usually much longer) proton spin–lattice relaxation time $T_{1\varrho}^H$ in the rotating frame, i.e. the time-constant for the decay of proton magnetization during the polarization process. Neglecting the small influence of carbon polarization on the proton magnetization it follows [50, 23, 294]

$$ M_C = M_H^0 \, \frac{T_{1\varrho}^H}{T_{1\varrho}^H - T_{CH}} \left[\exp\left(-\frac{t}{T_{1\varrho}^H} \right) - \exp\left(-\frac{t}{T_{CH}} \right) \right]. \tag{50} $$

Two interesting conclusions following from Eq. (50) are:

(i) Already theoretically a reduction of the optimal enhancement $M_H^0/M_C^0 = \gamma_H/\gamma_C \approx 4$ takes place.

With the abbreviations $\quad x = \dfrac{T_{CH}}{T_{1\varrho}^H} \quad$ and $\quad a = \dfrac{t_C}{t_C^{opt}},$

where t_C is the contact time and t_C^{opt} that for a maximal carbon magnetization according to Eq. (50), one gets

$$ M_C(t_C) = M_H^0 \cdot x^{\frac{ax}{1-x}} \cdot \frac{1-x^a}{1-x}. \tag{51a} $$

The reducing factor differs for small x-values ($T_{CH} \ll T_{1\varrho}^H$) only a little from 1. However, it decreases for derivations of the optimal contact time ($a \neq 1$). Its value at $x = 1$ ($T_{1\varrho}^H = T_{CH}$) is $a \cdot e^{-a}$.

(ii) Several carbon nuclei can be polarized in a different manner due to the different relaxation times. That is the case e.g. for the nonprotonated carbons with regard to the longer T_{CH}-values.

Even the contact times vary for different types of nuclei according to

$$ t_C^{opt} = \frac{T_{1\varrho}^H}{y - 1} \ln y, \tag{51b} $$

where $y = 1/x$ with x from Eq. (51a).

Two other features of this method yield an increase of sensitivity. Firstly, due to the very small effect of one Hartmann–Hahn contact on the proton spins (since the ^{13}C spins are rare) many repetitions of the cycle consisting of contact and observation can be done in principle, while protons are polarized only once. Unfortunately, for polymer applications one must often use only one contact for several reasons. Secondly and mainly, the repetition time for one experiment (which is very important for the maximal possible number of signal accumulations per time unit) is determined by the longitudinal relaxation time of the protons and not by the much longer relaxation time of the rare ^{13}C nuclei. In this way one gains an additional sensitivity enhancement by the factor $(T_{1C}/T_{1H})^{1/2}$. Because of the rare ^{13}C-spins and the proton decoupling one gets after Fourier transformation spectra without dipolar broadening (cf. Sect. 1.1.4) but with full anisotropies of the chemical shift.

1.2.6.2 Nuclear Overhauser Effect

Considering a double resonance experiment in liquids, e.g. with irradiation of proton frequency for decoupling and observation of ^{13}C nuclei, we find an enhancement of the ^{13}C signal caused by the cross-relaxation. Saturation of the proton yields, via spin–lattice relaxation, an equilibrium excess of nuclei in the lower ^{13}C energy level with respect to a Boltzmann distribution, hence, more *rf*-energy will be absorbed and the signal increases. The solution of the equation for the longitudinal magnetization of protons and ^{13}C nuclei under saturation conditions for the protons realized by random noise irradiation and under steady-state conditions leads to the overall enhancement factor NOE (Nuclear Overhauser effect). The latter represents the ratio between the ^{13}C-signals with and without proton decoupling. For dipolar interaction it follows according to [51, 52]

$$\text{NOE} = 1 + \eta = 1 + \frac{\gamma_H}{\gamma_C} \cdot \frac{6J(\omega_H + \omega_C) - J(\omega_H - \omega_C)}{J(\omega_H - \omega_C) + 3J(\omega_C) + 6J(\omega_H + \omega_C)} \tag{52}$$

with $J(\omega)$ from Eqs. (69, 77). The influence of other relaxation mechanisms, characterized by a longitudinal relaxation time T_1^O in contrast to T_1^D for dipolar interaction, can be expressed in the extreme narrowing case by the equation

$$\eta = \frac{\gamma_H}{2\gamma_C} \frac{(T_1^D)^{-1}}{(T_1^D)^{-1} + (T_1^O)^{-1}}. \tag{53}$$

Equation (52) shows the decay of the NOE-factor from a value of 2.99 under the extreme narrowing conditions to 1.15 for $\omega_C \tau_C \gg 1$, if only the dipolar interaction is effective. Deviations of the maximal value for the Overhauser enhancement can be caused by the anisotropy of the motion, e.g. by additional hindered internal motions [52, 53], as well as by the influence of other interactions according to Eq. (53). Therefore the NOE determinations support the relaxation time measurements to characterize the types of interactions and motional behaviour.

On the other hand different NOE enhancements for carbon within the same molecule are possible and one has to be careful with the determination of intensities in decoupling experiments.

1.3 Effect of Molecular Motion

One of the most important properties of the NMR-method is its sensitivity to the types of molecular motions. The very first works on NMR laid the basis of the theory relating magnetic relaxation parameters to microscopic characteristics of a substance. In [54–56] is shown that with considerably rapid motions the magnetic relaxation rates (T_1, $T_{1\varrho}$, and T_2) in the case of stochastic thermal fluctuations of some interactions (dipolar, quadrupolar, etc.) are proportional to the product of the interaction constant and a function, depending on some combinations of spectral densities $J(\omega)$ which are connected with a correlation function $\phi(t)$ by a Fourier transformation:

$$J(\omega) = \int\limits_{-\infty}^{+\infty} \phi(t/\tau)\exp(-i\omega\tau)\,d\tau, \tag{54}$$

where τ is the correlation time of the motion.

The function $\phi(t/\tau)$, is reduced to zero if t tends to infinity. In [57] is demonstrated that the spectral density formalism of the spin–lattice relaxation can also be applied to slower motions.

1.3.1 General Remarks on the Form of the Correlation Function

Let us divide the motion into two types: isotropic and anisotropic. We assume that isotropic motions are described by the one-parameter correlation function averaging the interaction A to zero.

The motion is considered anisotropic when its correlation function splits into two different time scales and can be described by two characteristic times τ_1 and τ_2 $(\tau_2 \gg \tau_1)$. In the particular case of spectral densities being determined in the frequency range, where the conditions $\omega\tau_1 \approx 1$, $\omega\tau_2 \gg 1$ are fulfilled, the correlation function of anisotropic motions can be written as

$$\phi(t/\tau) = A[(1 - q^2) + q^2 \phi'(t/\tau)], \tag{55}$$

where $\phi'(t/\tau)$ averages the interaction part $q^2 A$ to zero. The parameter q^2 can be the measure of motional anisotropy (amplitude factor). When $q^2 = 1$, the motion is isotropic; the parameter q^2 decreases with the increase of anisotropy.

In many cases, well known from dielectrical and mechanical measurements, the correlation function may be non-exponential and can often be determined by a distribution function $\varrho(\tau, \tau_0^h)$ of simple exponents (all distribution functions used here are normalized)

$$\phi'(t/\tau_0^h) = \int\limits_{0}^{\infty} \varrho(\tau, \tau_0^h)\exp(-t/\tau)\,d\tau \tag{56}$$

that is equivalent to an effective (homogeneous) spectrum of correlation times in the system. (τ_0^h is the parameter of the distribution and in symmetric cases the most probable correlation time).

When an intrinsic (inhomogeneous) distribution $G(\tau, \tau_0^i)$ of the correlation times due to the heterogeneity of the sample occurs, the function $\phi(t/\tau_0^i)$ is the following integral:

$$\phi(t/\tau_0^i) = \int\limits_{0}^{\infty} G(\tau, \tau_0^i)\phi(t/\tau)\,d\tau$$

$$= A \int\limits_{0}^{\infty} G(\tau_0^h, \tau_0^i)[(1 - q^2) + q^2 \phi'(t/\tau_0^h)]\,d\tau_0^h$$

$$= A \int\limits_{0}^{\infty} G(\tau_0^h, \tau_0^i)\left[(1 - q^2) + q^2 \int\limits_{0}^{\infty} \varrho(\tau, \tau_0^h)e^{-t/\tau}\,d\tau\right]d\tau_0^h. \tag{57}$$

It should be noted that, if $a = 0$ and ϕ' is the exponential correlation function, Eq. (57) becomes

$$\phi(t/\tau_0^i) \sim \int_0^\infty G(\tau, \tau_0^i) \exp(-t/\tau)\, d\tau \tag{58}$$

similar to Eq. (56).

If the nuclei take part in several (k) independent motions simultaneously the general correlation function can be represented as the product of correlation functions of these motions (ϕ_K):

$$\phi(t/\tau) = \prod_k \phi_k(t/\tau). \tag{59}$$

In the particular, but wide-spread case when the correlation times of the motions K are much greater than those of the motions $k-1$, i.e. $\tau_k \gg \tau_{k-1}$, the product may be transformed into a sum:

$$\phi_S(t/\tau) = \sum_k \phi_k(t/\tau). \tag{60}$$

As an example let us consider the case of two independent motions with the correlation functions of Eq. (55), averaging dipolar interaction. The general correlation function would be the product

$$\phi(t/\tau) = [a_1 + b_1\, \phi_1'(t/\tau_1)][a_2 + b_2\, \phi_2'(t/\tau_2)], \tag{61}$$

where $(a_1 + b_1)(a_2 + b_2) = \langle \Delta\omega^2 \rangle$ is the second moment, $a_1 a_2$ is the second moment part which does not depend on time, $b_1 a_2 + a_1 b_2 + b_1 b_2$ is the part depending on time. Here $a_2 b_1$ can be averaged only by the first motion, $a_1 b_2$ only by the second one and $b_1 b_2$ – by both motions.

If the condition $\tau_2 \gg \tau_1$ prevails, then $\phi_2'(t) \approx 1$ during the time of $\phi_1'(t)$ decay, Eq. (61) becomes

$$\phi(t/\tau) = \phi_S(t/\tau) = a_1 a_2 + b_1(a_2 + b_2)\phi_1'(t/\tau_1) + a_1 b_2 \phi_2'(t/\tau_2). \tag{62}$$

This example illustrates that the anisotropy of the more rapid motion is a necessary condition for the experimental observation of several motions in which the same nucleus takes part, the so-called multiplicity of relaxation transitions. This can be explained by the fact that the anisotropic motion does not reduce the interaction to zero, the slow motion averages the remaining part. Such process may continue till the next motion becomes isotropic and reduces the remaining part of the interaction to zero. The establishment of correlation functions for each kind of motion or at least the definition of some of their parameters is the desired aim of dynamic experiments. It should be noted that many dynamic methods with satisfactory sensitivity to the form of the correlation function cannot distinguish homogeneous and inhomogeneous distributions. The NMR-method as shown in [58] is a fortunate exception due to its ability to distinguish these two kinds of distributions in some experiments.

1.3.2 Two Approaches to the Analysis of Motional Effects

Quantitative information on molecular motion parameters can be obtained from the experimental data by two approaches that are called model and model-free (formal) [58–63].

The first approach consists in the following:

(i) Proceeding from physical considerations the model of motion is chosen.
(ii) The correlation function is derived for this model.
(iii) The magnetic relaxation parameters are calculated.
(iv) The comparison with the experimental data is made.

The second approach is carried out in another way:

(i) Formal expressions for NMR relaxation parameters are obtained by means of general correlation functions (56) or (57).
(ii) The characteristic behaviour of NMR signals and their parameters depending on the type of the correlation function are analysed.
(iii) The comparison between the experimental and the theoretical data is made. As a result, the type and the parameters of the correlation function corresponding to the experiment are determined.
(iv) The comparison of the obtained microdynamic characteristics with the correlation function parameters of different models is carried out.

The model approach, being direct and traditional, is limited, however, by the ambiguity and mathematical incorrectness in the analysis of such complex systems as polymers and biopolymers.

The first limitation is connected with the fact that very often several models can simultaneously satisfy the whole set of experimental data. Hence, there is a danger of ascribing to the investigated process a non-existing type of motion and to omit the real one.

The second one is due to the fact that the exact expressions for complicated models have more microdynamic parameters than experimentally calculated independent values. Therefore the problem can be solved only, in certain approximations, and not very elegantly.

Besides these two drawbacks a number of less important difficulties hinder the application of the model approach to the analysis of polymer systems. They are, for example, the complicated calculations of various NMR-signal parameters that hamper comparative analysis of the multi-parameter experiment for several models and several relaxation transitions simultaneously.

On the other hand, the model-free approach, based on a well worked out phenomenological foundation, is simple and practical. It allows, although at the expense of detailed information, the quick determination of the correlation function type and its parameters simultaneously for several transitions observed in the experiment within the framework of the same assumptions.

It can be supposed that the second approach is useful and effective at the first stage of investigation as it gives wide, though rough information on all the types of motion. This information may be the starting point for the construction of the more complicated model theory appropriate for more detailed analysis.

1.3.3 Model-free Approach

1.3.3.1 Expressions for the Signal Shape in Pulse NMR

In this section we shall introduce expressions for the shape of pulse NMR-signals and their parameters obtained for dipolar interaction of equal spins. The structure of the expressions of the magnetization for other types of interactions or for the dipolar interaction of unlike spins is similar, so there is no need to consider it here. We shall cite these interactions, if necessary, referring to the original papers.

Let us assume that the spin system of the sample is characterized by a mean local field with an instantaneous distribution described by a Gauss function. The averaging of the local field occurs due to the participation of nuclei in several independent types of motions and is described by the general correlation function Eq. (59). This means that the local field is divided into a series of components which are reduced to zero by corresponding motions. Each motion can be described both by a non-exponential correlation function and by a inhomogeneous distribution of correlation times. Assuming that $\phi_S(t)$ is a sum we neglect cross-terms of the local field that are averaged by several motions simultaneously.

We can write the general expressions for longitudinal and transverse decays using the well-known results of the NMR theory (cf. e.g. [1])

$$A_{1,2,1\varrho}(t) = \sum_{k=1}^{n} \int_0^\infty G_k(\tau_0^h, \tau_{0k}^i) \exp\left\{ -t \int_0^\infty \frac{\varrho_k(\tau, \tau_0^h)}{T_{1,2,1\varrho}^k(\tau)} d\tau \right\} d\tau_0^h, \tag{63}$$

where

$$\frac{1}{T_1^k} = \tfrac{2}{3} C_k [J^k(\omega_0) + 4J^k(2\omega_0)], \tag{64}$$

$$\frac{1}{T_{1\varrho}^k} = \tfrac{1}{3} C_k [3J^k(2\omega_e) + 5J^k(\omega_0) + 2J^k(2\omega_0)], \tag{65}$$

$$\frac{1}{T_2^k} = \tfrac{1}{3} C_k [3J^k(0) + 5J^k(\omega_0) + 2J^k(2\omega_0)] \quad \text{(if } \omega_0\tau_0 \lesssim 1), \tag{66}$$

$$\frac{1}{T_2^k} = C_k \frac{\tau^2}{t} f\left(\frac{t}{\tau}\right) \quad \text{(otherwise)} \tag{67}$$

with

$$C_k = q_k^2 \langle \Delta\omega^2 \rangle_{k-1}, \tag{68}$$

$$J^k(\omega) = \frac{\tau}{1 + \omega^2\tau^2}, \tag{69}$$

$$f\left(\frac{t}{\tau}\right) = e^{-t/\tau} - 1 + \frac{t}{\tau}, \tag{70}$$

$$\omega_e^2 = \omega_1^2 + \omega_{\text{loc}}^2. \tag{71}$$

$G(\tau_0^h, \tau_0^i)$ is the distribution of correlation times due to the heterogeneity and $\varrho(\tau, \tau_0^h)$ due to the nonexponentiality of the correlation functions. The form of Eq. (67) was chosen for a better agreement with the preceding equations, although a relaxation time in this case does not exist.

The motion of the type k is by definition slower than that for $(k-1)$. $\langle \Delta\omega^2 \rangle_{k-1}$ represents that part of the second moment which remains after averaging by all faster types of motion, i.e. $\langle \Delta\omega^2 \rangle_0$ corresponds to the general second moment in absence of motion. Then the anisotropy parameter q_k^2 is defined as the part of the second moment which is averaged by the type k of motion related to the remaining second moment after all faster motions:

$$q_k^2 = \frac{\langle \Delta\omega^2 \rangle_{k-1} - \langle \Delta\omega^2 \rangle_k}{\langle \Delta\omega^2 \rangle_{k-1}} \tag{72}$$

and consequently,

$$C_k = \langle \Delta\omega^2 \rangle_{k-1} - \langle \Delta\omega^2 \rangle_k. \tag{73}$$

In Fig. 4 the temperature dependence of the second moment is schematically shown for two different types of anisotropic and one isotropic motion. In the temperature regions near T_1, T_2 and T_3, where the conditions $\tau_k(\langle \Delta\omega^2 \rangle_{k-1} - \langle \Delta\omega^2 \rangle_k)^{1/2} \approx 1$ are fulfilled, jumps of the second moment are shown which are quantitatively described in Eq. (72).

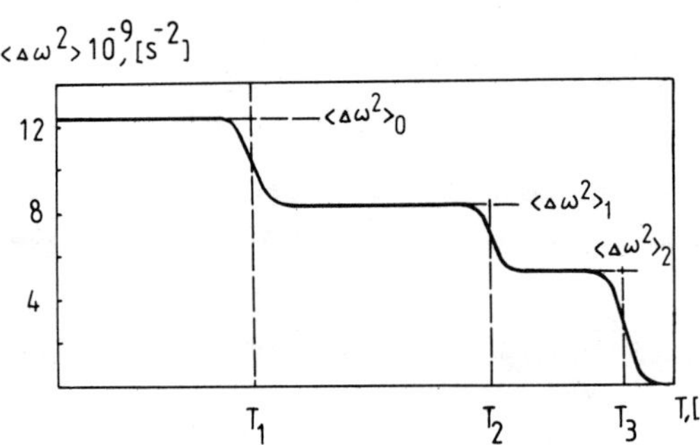

Fig. 4. Temperature dependence of the second moment for polymers which underlie two anisotropic and one isotropic motions ($T_{1,2,3}$ are the temperatures for the corresponding relaxation transitions)

The formulas describing the magnetization decay for the motion number k in some important particular cases can be obtained from the general Eq. (63).

So, without inhomogeneous distribution of correlation times we have

$$A_{1,2,1\varrho}^k(t, \tau_{0k}^h) = \exp\left[-t \int_0^\infty \frac{\varrho_k(\tau, \tau_{0k}^h)}{T_{1,2,1\varrho}(\tau)} d\tau \right] \tag{74}$$

while inhomogeneous distribution of the exponential correlation functions transforms these expressions into

$$A_{1,2,1\varrho}^k(t, \tau_{0k}^i) = \int_0^\infty G_k(\tau, \tau_{0k}^i) \exp\left[-\frac{t}{T_{1,2,1\varrho}(\tau)} \right] d\tau. \tag{75}$$

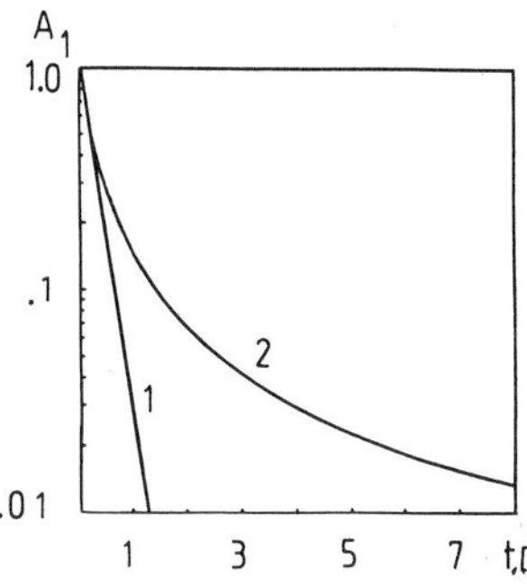

Fig. 5. Transversal magnetization decay for homogeneous (*1*) and inhomogeneous (*2*) distributions of correlation times. $G(\tau)$ is described by Eq. (80) with $\alpha = \beta = 0.5$, $\tau_0 = 2 \times 10^{-10}$ s

A principal difference in the behaviour of these Eqs. (74) and (75) is shown. According to Eq. (74) the longitudinal magnetization decay is always described by simple exponents and the transverse magnetization decay transforms from Gaussian to exponential, while according to Eq. (75) the longitudinal magnetization decay is a sum of exponents (see Fig. 5) and the decay of transversal magnetization has an even more complicated shape. Such a magnetization behaviour makes it principally possible to distinguish in the experiment correlation time spectra of different natures.

It should be noted that when inhomogeneous distribution of correlation times takes place transitions between the states with different correlation times can be observed. These transitions are caused by material diffusion as well as by spin diffusion processes. The condition of rapid exchange is considered to be fulfilled when the intergroup (intermolecular) dipolar interaction is large enough to secure an effective magnetization transfer between the relaxators during the relaxation time or when the life-time of each state is much shorter than the correlation time. The averaged signals can be observed in the experiment. Magnetization decays in this case will be described by expressions identical with Eq. (74) and the case cannot be distinguished from homogeneous distribution. An intermediate case, when the condition of rapid exchange is fulfilled only for some relaxators with broad distributions and not for the others is considered in [64].

1.3.3.2 Expressions for Nuclear Magnetic Relaxation Times

Expressions for relaxation times depending on the distribution parameters can easily be obtained from Eq. (74) for homogeneous distribution (nonexponential correlation function):

$$\frac{1}{T_{1,2,1\varrho}} = \sum_k \frac{1}{T^k_{1,2,1\varrho}}, \tag{76}$$

where T_1, T_2, $T_{1\varrho}$ follow from Eqs. (64–67), however $J(\omega)$ has to be replaced by $J^k(\omega)$ according to

$$J^k(\omega) = \int_0^\infty \frac{\varrho_k(\tau, \tau^h_{0k})\tau}{1 + (\omega\tau)^2} d\tau. \tag{77}$$

For the transversal relaxation time T_2 instead of Eq. (66), which is valid for $\langle \Delta\omega^2 \rangle_{k-1} \tau^2 \ll 1$ (high-frequency limit), for slower motions $\langle \Delta\omega^2 \rangle_{k-1} \tau^2 \approx 1$ an approximately valid expression, derived for the dependence of the second moment on τ [1], can be used slightly modified:

$$\langle \Delta\omega^2 \rangle_{k-1} - \langle \Delta\omega^2 \rangle_k = \left(\frac{1}{T_2^{k-1}} \right)^2 = C_k \frac{2}{\pi} \int_0^\infty \varrho(\tau, \tau_{0k}^h) \arctan\left(\frac{\pi}{2} \frac{\tau}{T_2^{k-1}(\tau)} \right) d\tau. \tag{78}$$

The concept of a relaxation time becomes indefinite for inhomogeneous distribution (Eq. (75)) since the magnetization decays cannot be described by a single time constant. This case makes it necessary to define a relaxation time either as an averaged value or as the time in which the magnetization decreases to the value e^{-1}.

It should be noted that the longitudinal magnetization decays due to a rapid exchange (material or spin diffusion) are often exponential, even under inhomogeneous distribution conditions, and Eqs. (64, 77) can always be applied to them, the information on the nature of the spectrum is lost however.

The influence of various distributions on nuclear relaxation time is studied in detail in the literature [65–69]. A summary of typical distribution functions with their influence on the characteristic behaviour of magnetic relaxation times is given in the review article by McBrierty [70]. All the distribution functions used for dielectric as well as for NMR investigations, characterize, in a special manner, the position and the width of the distributions and, when required, their asymmetry. The Fuoss–Kirkwood (FK) distribution [71] especially yields simple expressions for the spectral density, which are very similar to the distribution given for a single correlation time process by Eq. (69). Therefore we have derived a modification of the Fuoss–Kirkwood function [72] which also permits us to describe asymmetric relaxation time dependence on the temperature. Thereby we consider the temperature dependence on the correlation time, as it is usually done, by an Arrhenius process:

$$\tau_c = \tau_c^0 \exp\left[\frac{E_a}{kT} \right], \tag{79}$$

where τ_c^0 is the correlation time for infinite temperatures, E_a the activation energy under consideration, and k the Boltzmann constant.

The normalized modified distribution function and the corresponding spectral density are given by

$$G(\tau, \tau_0) = \frac{\alpha+\beta}{\pi\tau} \sin\left(\frac{\alpha\pi}{\alpha+\beta} \right) \frac{\left(\frac{\tau}{\tau_0} \right)^\beta \cos\left(\alpha \frac{\pi}{2} \right) + \left(\frac{\tau}{\tau_0} \right)^{-\alpha} \cos\left(\beta \frac{\pi}{2} \right)}{\left(\frac{\tau}{\tau_0} \right)^{\alpha+\beta} + \left(\frac{\tau}{\tau_0} \right)^{-(\alpha+\beta)} + 2\cos\left[(\alpha+\beta) \frac{\pi}{2} \right]}, \tag{80}$$

$$J(\omega\tau_0) = \int_0^\infty G(\tau, \tau_0) \frac{\tau \, d\tau}{1+(\omega\tau)^2} = \frac{\alpha+\beta}{2\omega} \sin\left(\frac{\alpha\pi}{\alpha+\beta} \right) \frac{(\omega\tau_0)^\alpha}{1+(\omega\tau_0)^{\alpha+\beta}}. \tag{81}$$

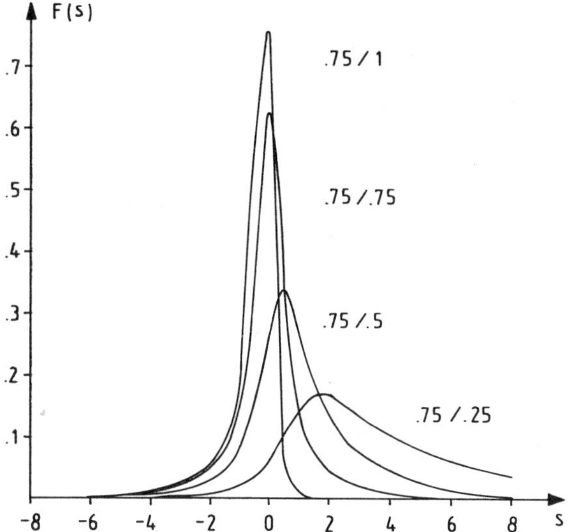

Fig. 6. Normalized distribution function $F(S)$ according to Eqs. (80, 82). Parameters on the curves: α/β. τ_0 and S have been modified according to Eq. (83a) in such a way that $(\omega\tau_0')_{\max} = 1$

In Fig. 6 the normalized distribution function $F(S)$ according to

$$S = \ln\frac{\tau}{\tau_0} \tag{82a}$$

and

$$F(S) = \tau\, G(\tau) \tag{82b}$$

is shown.

Figure 7 represents the temperature and frequency dependences of T_1, T_2 and $T_{1\varrho}$ for several width parameters.

Unfortunately, for the calculation of $A_2(t)$ according to Eq. (74) using Eqs. (66) and (81) the integral diverges for the term $J(0)$ (if $\alpha \neq 1$), as is also the case for many other distributions. Therefore, for this an equation similar to Eq. (78) was used.

The main advantage of the distribution given by Eqs. (80, 81) is the form of the spectral density $J(\omega\tau_0)$. The relatively simple expression is again similar to that of Eq. (69) and all possible slopes of $\log J^{-1}(\omega\tau_0)$ versus $\log \tau/\tau_0$ (except for a constant factor also versus $1/T$) have in the high and the low temperature regions the values $-\alpha$ and $+\beta$, respectively, where $0 < \alpha, \beta \leq 1$. This explanation is possible for all the values of α and β, especially for $\alpha = \beta$, $\alpha = 1$, or $\beta = 1$, giving the special cases of a FK-distribution and two others equivalent to those of Cole–Davidson [73] and Fang [74], respectively, without any analytical change of the functions $G(\tau, \tau_0)$ and $J(\omega\tau_0)$.

Furthermore, the value $(\omega\tau_0)_{\max}$ for a maximal spectral density $J_{\max}$ can be given analytically

$$(\omega\tau_0)_{\max} = \left(\frac{\alpha}{\beta}\right)^{\frac{1}{\alpha+\beta}}, \tag{83a}$$

$$J_{\max} = \frac{\beta}{2\omega}\sin\left(\frac{\alpha\pi}{\alpha+\beta}\right)\left(\frac{\alpha}{\beta}\right)^{\frac{\alpha}{\alpha+\beta}}. \tag{83b}$$

Thus, the modified Fuoss–Kirkwood distribution permits (i) from both slopes of the temperature dependence of J, the determination of the width parameters α and β (for the known activation energy), (ii) the destination of the mean correlation time and of the second moment from the minimum ((i) and (ii) without computer fitting), and (iii) computer fitting of the relaxation curves with a common analytical expression for all the cases (e.g. also for symmetric curves, where a FK-distribution follows automatically).

In the case of quadrupolar interaction (spin $I = 1$) expressions such as Eqs. (64, 66) hold true for isotropic motion, where [270]

$$C_k = \frac{g}{80}\left(\frac{e^2 q Q}{\hbar}\right)^2\left(1 + \frac{\eta}{3}\right). \tag{83c}$$

(Explanation of symbols cf. Sect. 1.1.3.)

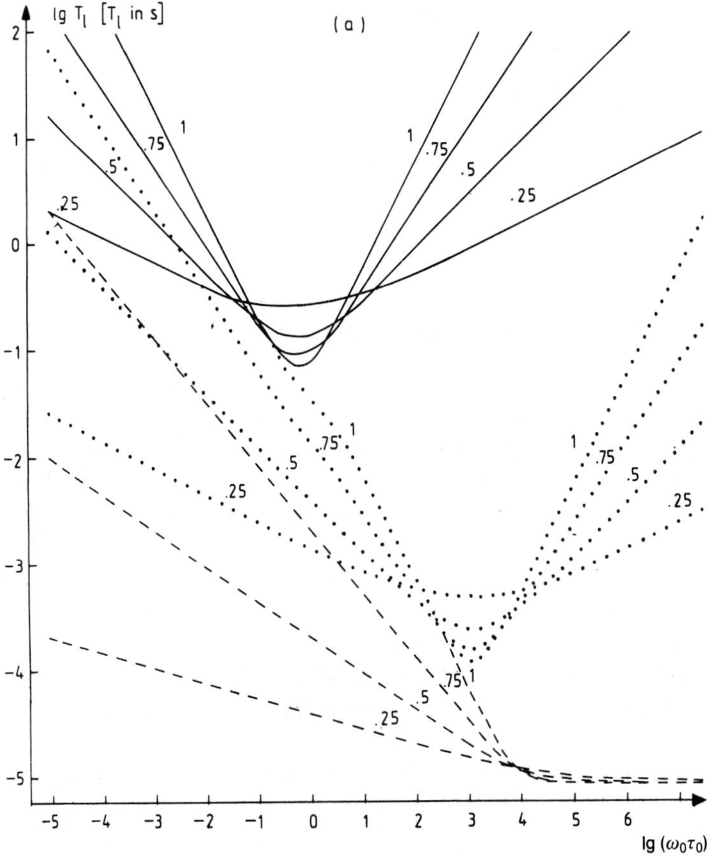

Fig. 7a.

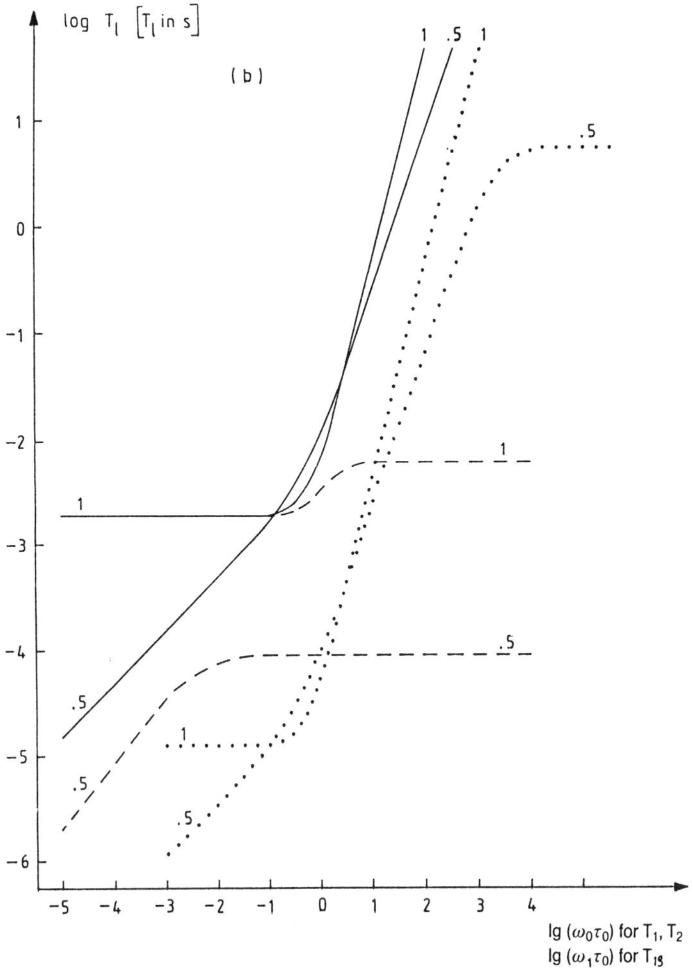

Fig. 7. Temperature (**a**) and frequency (**b**) dependences of T_1, T_2, $T_{1\varrho}$ for distribution functions $G(\tau)$ according to Eq. (80). Full lines: T_1, dashed lines: T_2, dotted lines: $T_{1\varrho}$. Parameters on the curves: For T_1 and $T_{1\varrho}$ at $\omega\tau \ll 1$: α and at $\omega\tau \gg 1$: β; for T_2:α. (T_2 is nearly independent on β; for T_1 and $T_{1\varrho}$ the curves for $\alpha \neq \beta$ accept in the region $\omega\tau \approx 1$ approximately values corresponding to the mean value of α and β). Concerning τ_0 see Fig. 6

For the quadrupolar relaxation time of spin alignment T_{1Q} follows

$$T_{1Q}^{-1} = 2C_k J(\omega_0). \tag{83d}$$

Thus, T_{1Q} is of the same order as T_1. The ratio T_{1Q}/T_1 changes monotonously from 5/3 in the extreme narrowing case up to 2/3 in the limit of slow motion.

The influence of anisotropic types of motion on the spin–lattice relaxation in solids was shown in [283].

Concerning the influence of correlation time distributions on the relaxation times analogous treatment like that described above is possible for quadrupolar interaction too.

1.3.4 Correlation Functions for Some Motional Models

In a review article [70] McBrierty gives a critical and detailed consideration to a great number of various models that are used in the analysis of polymer NMR-experiments. Without repeating this analysis let us consider only the most simple models which are the basis for the more complicated ones. These motional models are sufficient to interpret data of solid polymers, while more complicated special models based on motional characteristics of polymer molecules are used only in the analysis of melts and solutions.

The Rouse-model [75], the De Gennes model of reptations [76] and a model taking into account the entanglements [77] are such examples. With the application to NMR experimental analysis these models are discussed in detail in [70, 78–83].

1.3.4.1 Simple Motional Models

Some simple types of anisotropic motion can be described by a general correlation function of the type of Eq. (55):

$$\phi(t/\tau) = A[(1-q^2) + q^2 e^{-t/\tau}], \tag{84}$$

and the corresponding spectral density is

$$J(\omega) = \frac{q^2 A\tau}{1+(\omega\tau)^2}, \tag{85}$$

where the anisotropy parameter $q^2 \leq 1$ ($q^2 = 1$ means isotropic motion) can be calculated using an expression given in [33]:

$$q^2 = \tfrac{3}{2}[1 - \overline{\cos^2\alpha(t)}]. \tag{86}$$

The time averaging has to be performed as to the angle $\alpha(t)$ of the internuclear vector under consideration with respect to its direction at $t = 0$. From Eq. (86) values for q^2 can be derived, for special models, e.g. (for arbitrary starting positions) [84] (cf. Fig. 8):

(i) Jumps about an angle δ between two equivalent sites:

$$q^2 = \tfrac{3}{4}\sin^2\delta \tag{87}$$

(ii) Jumps about angles δ between three sites at the corners of an equilateral triangle:

$$q^2 = \sin^2\delta \tag{88}$$

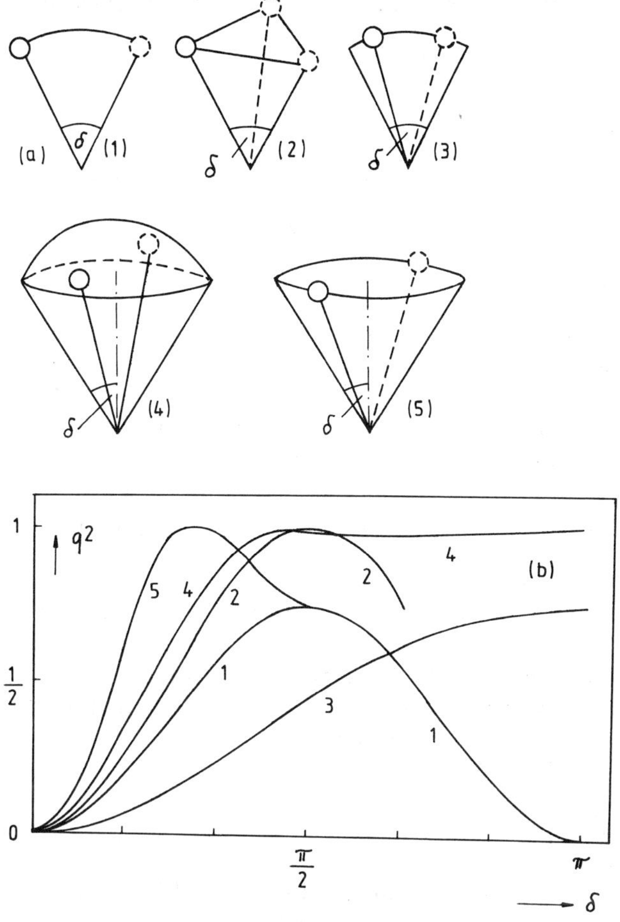

Fig. 8. Dependence of the anisotropy parameter on the characteristic angle (**b**) for several models (**a**) of rotation–oscillation motions according to Eqs. (87–91)

(iii) Continuous sites at an arc with the opening angle δ:

$$q^2 = \frac{3}{4}\left[1 - \left(\frac{\sin\delta}{\delta}\right)^2\right].\tag{89}$$

This result agrees with that derived from the expressions given in [85].

(iv) Continuous sites inside a cone with the half opening angle δ:

$$q^2 = 1 - \left[\frac{\cos\delta}{2}(1+\cos\delta)\right]^2\tag{90}$$

(v) Continuous sites at a circle on the sphere with the half opening angle δ (in accordance with a rotation around a fixed axis [86]):

$$q^2 = 3\sin^2\delta(1 - \tfrac{3}{4}\sin^2\delta).\tag{91}$$

(The same result follows for each arbitrary number ≥ 3 of equidistant sites along that circle).

1.3.4.2 Defect–Diffusion Models

The correlation function for a defect diffusion model was derived by Glarum [87]:

$$\phi\left(\frac{t}{\tau_{1,d}}\right) = \exp\left(\frac{t}{\tau_1}\right)\mathrm{erfc}\left(\frac{t}{\tau_d}\right)^{1/2}, \tag{92}$$

where τ_1 is the correlation time of the dipole motion without defects and τ_d is the time in which the defect passes the average distance between two neighbouring defects.

A great number of various modifications of this model is discussed in the literature [34, 88–92], especially by the groups of Kimmich [88–90] and Skolnik [92].

The analysis of Eq. (92) demonstrated three important cases of simplification of the form of this function:

(i) $\tau_1 \ll \tau_d$, the contribution of the diffusion term can be neglected, and Eq. (92) turns into a simple exponent.
(ii) $\tau_1 \approx \tau_d$, Eq. (92) is reduced to a function corresponding to the asymmetric Cole–Davidson distribution with the parameter $\beta = 0.5$ or to the modified FK-distribution with $\alpha = 1$ and $\beta = 0.5$, respectively.
(iii) $\tau_1 \gg \tau_d$, the contribution of the diffusion term dominates and Eq. (92) turns into a function corresponding to the symmetric Cole–Cole [73] or FK distributions with $\beta = 0.5$.

In an analogous manner a comparison of several diffusion models [91, 93] showed for the τ- as well as for the ω-dependences of the spectral density in all cases $\omega\tau \gg 1$, ≈ 1, $\ll 1$ the same behaviour as for a FK- or a Cole–Cole-distribution with $\beta = 0.5$.

Literature on the dielectric relaxation deals with a great number of motional models with nonexponential correlation functions. Many of these models were described by the so-called Williams–Watts function, suggested earlier by Kohlrausch in the 19th century

$$\phi\left(\frac{t}{\tau}\right) = \exp\left[-\left(\frac{t}{\tau}\right)^{\delta}\right]. \quad (0 < \delta \leq 1) \tag{93}$$

In [94, 95] the spectral behaviour of this function is shown to be similar to that of the well-known Cole–Davidson distribution [73]. Papers by Shore and Zwanzig [96] demonstrate that the model of monomeric coupled oscillators lead in all cases to the same correlation function with the value of the parameter varying within the interval 0.5 ... 1.

Let us give the correlation function obtained by Kimmich for De Gennes reptation model [79] as an example for the construction of the correlation function for a complex motion of a polymer chain.

In Kimmich's interpretation this function is presented in the following way:

$$\phi(t) = \phi_a(t)\phi_b(t)\phi_c(t), \tag{94}$$

where

$$\phi_a(t) = a_1 \exp\left(-\frac{t}{\tau_s}\right) + a_2. \tag{95}$$

The component "a" is due to anisotropic local motions such as short distance diffusion of defects, τ_s is an effective time constant characterizing the local motions.

$$\phi_b(t) = \exp\left(\frac{t}{2\tau_l}\right)\mathrm{erfc}\left(\frac{t}{2\tau_l}\right)^{1/2}. \tag{96}$$

The component "b" is the reptation along the tube bends with $\tau_l = l^2/2D_l$, τ_l is the mean diffusion time over the correlation length l of the tube orientation. Here the chain is assumed to diffuse with a constant diffusion coefficient D_l within a stationary tube of infinite length.

The component "c" or tube renewal leads to the definite loss of correlation with the initial segment orientation. Two mechanisms corresponding to the end and the middle chain sections contribute to this component. Both mechanisms result in an exponential correlation function with the correlation time τ_r.

1.3.5 Quadrupolar Line Shape under the Influence of Molecular Motion

1.3.5.1 Timescale of Motion

The relaxation time behaviour discussed in the previous sections yields information concerning the correlation times as well as the special motional model. In order to determine how the interpretation of deuteron spectra can give the analogous information, we shall discuss the example of the chemical exchange of two species with resonance frequencies split in absence of motion by $\pm\Delta$ relative to the mean frequency which is taken as the origin on the frequency scale. With an exchange rate Ω it follows for the line shape [1]

$$I(\omega) \propto \frac{4\Delta^2\Omega}{(\omega^2 - \Delta^2)^2 + (2\omega\Omega)^2}. \tag{97}$$

Eq. (97) yields [16]

— in the fast exchange limit ($\Omega^2 \gg \Delta^2$) a Lorentzian at $\omega = 0$ with a linewidth (full width at half height) Δ^2/Ω,
— in the intermediate exchange region ($\Omega^2 \approx \Delta^2$), e.g. for $\Omega^2 = \Delta^2/2$ a Lorentzlike line at $\omega = 0$ with a linewidth 2Δ, i.e. the signal is spread even further than the frequency range,
— in the slow exchange limit ($\Omega^2 \ll \Delta^2$) two Lorentzians at $\omega = \pm\Delta$ with a linewidth 2Ω.

With regard to the natural linewidth of the two initial lines assuming Lorentzians with a linewidth $2/T_2^*$ (T_2^* being the effective transversal relaxation time including broadening due to the field inhomogeneity), we have to add the value $2/T_2^*$ in all cases to the full width at half height. This is of no influence in the intermediate exchange region, plays a role in the fast as well as in the slow exchange limit and is at last the only important contribution to the linewidth of the one line for ultrafast ($\Delta^2/\Omega \ll 2/T_2^*$) or of the two lines for ultraslow ($\Omega \ll 1/T_2^*$) exchange. Thus, the signal change by molecular motion is limited by $2/T_2^* \approx 4 \times 10^3\,\mathrm{s}^{-1}$ given by dipolar interaction and field inhomogeneities.

Under the assumptions used above that we deal with single crystal type spectra, line shape analysis of deuteron spectra should permit the determination of correlation times in a range from approximately 10^{-7} up to approx. 10^{-4} s [16]. For polymers, however, we measure powder spectra with an extent on the ω-scale of 4δ (see Fig. 2) with $\delta \approx 4 \times 10^5\,\mathrm{s}^{-1}$ which is now the limiting quantity for using line shape analysis. This restricts the dynamic range drastically by two orders of magnitude. Thanks to the basic work of Spiess and Sillescu [269, 270] it has been possible to circumvent these difficulties by introducing suitable pulse techniques enabling the investigation of slow and ultraslow motions in polymers. Before we briefly describe the main ideas of these echo methods, the possibilities for investigating fast motional processes should be explained.

1.3.5.2 Fast Motion

The line shape in absence of motion was derived from Eq. (38) and represented in Fig. 2. The powder pattern shows, for axial symmetry, two infinities at $\omega - \omega_0 = \mp\delta$ and two edges at $\pm 2\delta$, whereas in the general case two infinities occur at $\omega - \omega_0 = \mp(1-\eta)\delta$ and two pairs of edges at $\pm 2\delta$, and $\mp(1+\eta)\delta$ [11, 267]. In the case of fast motion, if the correlation frequency is large compared with the extent of the spectrum in absence of motion, it can be averaged over the motional process, i.e. the electric field gradient tensor (fgt) must be transformed into a new frame of the averaged fgt. Then Eq. (38) must be transformed in the averaged fgt frame replacing θ and ϕ with ϑ and φ, respectively; the polar angles specify the $\vec{B}_0$ orientation with respect to the principal axes system of the averaged fgt. The new equation contains averaged values $\bar{\delta}$ and $\bar{\eta}$, where $\bar{\eta}$ can be non zero although the initial tensors are usually axially symmetric for $C-^2H$ bonds [16, 267].

The (traceless) fgt $\boldsymbol{D}$ can therefore be written in the principal axes system in the following way:

$$\boldsymbol{D} = \delta \begin{pmatrix} -\frac{1}{2} & 0 & 0 \\ 0 & -\frac{1}{2} & 0 \\ 0 & 0 & 1 \end{pmatrix}. \tag{98}$$

Such an axially symmetric tensor is not changed for a rotation about the z axis. Thus, an arbitrary reorientation, usually expressed by three rotations about the Eulerian angles, can be expressed in this case by two rotations only. These are a first rotation of an angle β about the y axis followed by a second one of an angle α

about the new z axis. These transform $\boldsymbol{D}$ into $\boldsymbol{T}$:

$$\boldsymbol{T} = \boldsymbol{R}_z(\alpha)\boldsymbol{R}_y(\beta)\boldsymbol{D}[\boldsymbol{R}_z(\alpha)\boldsymbol{R}_y(\beta)]^T \tag{99}$$

with

$$\boldsymbol{R}_z(\alpha) = \begin{pmatrix} \cos\alpha & -\sin\alpha & 0 \\ \sin\alpha & \cos\alpha & 0 \\ 0 & 0 & 1 \end{pmatrix}, \quad \boldsymbol{R}_y(\beta) = \begin{pmatrix} \cos\beta & 0 & \sin\beta \\ 0 & 1 & 0 \\ -\sin\beta & 0 & \cos\beta \end{pmatrix}. \tag{100}$$

That yields

$$\boldsymbol{T} = \frac{3\delta}{2}\begin{pmatrix} \cos^2\alpha\sin^2\beta - \frac{1}{3} & \sin\alpha\cos\alpha\sin^2\beta & \cos\alpha\sin\beta\cos\beta \\ \sin\alpha\cos\alpha\sin^2\beta & \sin^2\alpha\sin^2\beta - \frac{1}{3} & \sin\alpha\sin\beta\cos\beta \\ \cos\alpha\sin\beta\cos\beta & \sin\alpha\sin\beta\cos\beta & \cos^2\beta - \frac{1}{3} \end{pmatrix}. \tag{101}$$

If all nondiagonal elements of $\boldsymbol{T}$ vanish according to the special motional process the averaged values $\bar{\delta}$ and $\bar{\eta}$ will be calculated from

$$\bar{\delta} = t_{zz}, \quad \bar{\eta} = \frac{t_{yy} - t_{xx}}{t_{zz}}, \tag{102}$$

where the choice of the sequence x, y, z takes place according to $|t_{zz}| \geq |t_{xx}| \geq |t_{yy}|$ (i.e. $0 \leq \eta \leq 1$).

They replace δ and η in Eq. (38) and can be used for a discussion of the line shape because the replacing of θ and ϕ by ϑ and φ does not influence the powder spectrum.

Let us consider some special cases where the nondiagonal elements of $\boldsymbol{T}$ vanish:

(i) $\alpha = 0$, β jumps between $-\beta_0$ and β_0 (or totally equivalent: $\beta = \beta_0$, α jumps between 0 and π):

Thus it follows that

$$t_{11} = \frac{\delta}{2}(3\sin^2\beta_0 - 1), \quad t_{22} = -\frac{\delta}{2}, \quad t_{33} = \frac{\delta}{2}(3\cos^2\beta_0 - 1). \tag{103}$$

For $45° \leq \beta_0 \leq 90°$ (and exactly for $(90° - \beta_0)$ also) we have two singularities at $\pm\delta(3\cos^2\beta_0 - 1)$ and two pairs of edges at $\mp\delta$ and $\pm\delta(3\sin^2\beta_0 - 1)$. Especially for a two-site hop within a kink-3-bond motion ($\beta_0 = \beta_m$, magic angle) follows $\bar{\delta} = \delta/2$, $\bar{\eta} = 1$, i.e. one singularity at $\omega - \omega_0 = 0$, two edges at $\pm\delta$ (see Fig. 2c).
For $180°$-phenyl-ring flips ($\beta_0 = 60°$) is $\bar{\delta} = \frac{5}{8}\delta$, $\bar{\eta} = \frac{3}{5}$, i.e. two singularities at $\omega - \omega_0 = \mp\delta/4$ and two pairs of edges at $\mp\delta$ and $\pm\frac{5}{4}\delta$, respectively (see Fig. 2b).

(ii) $\beta = \beta_0$, α diffuses between 0 and 2π (or totally equivalent: $\beta = \beta_0$, α jumps between α_0, $\alpha_0 + 120°$, $\alpha_0 - 120°$):

$$t_{11} = t_{22} = \frac{\delta}{4}(1 - 3\cos^2\beta_0), \quad t_{33} = -2t_{11}. \tag{104}$$

This gives for arbitrary β_0 values an axial symmetric pattern (see Fig. 2a) with two singularities at $\mp\delta(3\cos^2\beta_0 - 1)$ and two edges at $\pm\delta(3\cos^2\beta_0 - 1)$.

For a methyl group diffusing between stochastic positions around the symmetry axis or jumping between three equilibrium states ($\beta_0 = 2\beta_m$) it follows $\bar{\delta} = \delta/3$, $\bar{\eta} = 0$, i.e. two infinities at $\omega - \omega_0 = \pm \delta/3$ and two edges at $\mp 2\delta/3$.

For a free phenyl ring diffusion ($\beta_0 = 60°$) it follows $\bar{\delta} = -\delta/8$, $\bar{\eta} = 0$, i.e. two singularities at $\pm \delta/8$ and two edges at $\mp \delta/4$.

Therefore, it is usually possible to make a clear distinction between different models of motion. There are, however, cases such as diffusion between stochastic positions or jumping between three equilibrium states of a methyl group, in which it is impossible to distinguish between two possibilities just by the interpretation of the line shape. The same holds for two-site jumps between $+\beta_0$ and $-\beta_0$ or $+(90° - \beta_0)$ and $-(90° - \beta_0)$, respectively. Likewise, one can get exactly the same line shapes as in (i) by assuming again $\alpha = 0$, but β diffuses between $-\beta_0$ and $+\beta_0$, with somewhat changed angles β_0.

1.3.5.3 Slow Motion

The application of the solid-echo technique to the investigation of polymer samples has two main advantages. Firstly the disadvantages mentioned at the end of Sect. 1.3.5.1 have been circumvented and the formation of the solid echo is limited by T_2^* only. Secondly, independently of the rapid FID decay with corresponding dead-time problems, the signal acquisition can be started at the echo maximum giving an undistorted spectrum after the Fourier transformation [259].

To explain the influence of motion on the spectrum, we shall discuss here briefly the way for derivation of Eq. (97). If the motional process is described by a stationary Marcovian process for the jumps between different possible frequencies (e.g. caused by different angles of the unique fgt axis with respect to the external field $\vec{B}_0$), then the relaxation function for the FID can be expressed [1] by

$$G(t) = \vec{W} \cdot \exp\{(i\omega + \pi)t\} \cdot \vec{1}, \tag{105}$$

where

$\vec{1}$ is a vector with all components equal to unity,

$\vec{W}$ is a vector containing the initial probabilities for several frequencies, often equal a priori probabilities,

ω is a diagonal matrix representing the different possible frequencies and

π is a matrix elements of which characterize the jump rate between the frequencies ω_α and ω_β.

The absorption signal is then

$$I(\omega) = re \int_0^\infty G(t)e^{i\omega t}\,dt = re\,\vec{W} \cdot \int_0^\infty \exp\{i(\omega - \omega E)t + \pi t\}\,dt \cdot \vec{1} \tag{106}$$

$$= -re\,\vec{W} \cdot [i(\omega - \omega E) + \pi]^{-1} \cdot \vec{1}.$$

Under the assumptions of Eq. (97) an analytical solution is possible with

$$\vec{W} = \vec{1},\, \omega = \begin{pmatrix} \Delta & 0 \\ 0 & -\Delta \end{pmatrix},\quad \pi = \begin{pmatrix} -\Omega & \Omega \\ \Omega & -\Omega \end{pmatrix},\quad E = \begin{pmatrix} 1 & 0 \\ 0 & 1 \end{pmatrix} \tag{107}$$

giving exactly the expression for $I(\omega)$ represented in Eq. (97). With regard to the powder pattern and to the more complicated models, one has to add the following steps:

(i) Choice of the ω_i values according to $\omega/\omega_0 = 3\cos^2\theta - 1$ in agreement with the corresponding angles to the $\vec{B}_0$ field.

(ii) Solution of Eq. (106) by diagonalization [267, 269]:

$$\lambda = S^{-1}(i\omega + \pi)S, \tag{108}$$

$$I(\omega) = re\,\vec{W}\cdot S\cdot \int_0^\infty \exp\{(\lambda - i\omega E)t\}\,dt\cdot S^{-1}\cdot\vec{1}$$

$$= -re\,\vec{W}\cdot S\cdot(\lambda - i\omega E)^{-1}\cdot S^{-1}\cdot\vec{1}. \tag{109}$$

Inversion of the diagonal matrix $(\lambda - i\omega E)$ is now simple. Diagonalization must be carried out only once for each orientation, valid for an arbitrary number of ω values. Therefore, taking into account the computational expense, this treatment is preferable to the direct matrix inversion of $[i(\omega - \omega E) + \pi]$ in Eq. (106).

In an analogous manner the calculation of the solid-echo spectrum under the influence of motion can be performed. The relaxation function is now given by [269, 272] (τ being the pulse distance)

$$G(t, \tau) = \vec{W}\cdot\exp\{(i\omega + \pi)\tau\}\exp\{(-i\omega + \pi)(t - \tau)\}\cdot\vec{1}. \tag{110}$$

Fourier transformation has to be carried out with regard to $(t - 2\tau)$. Therefore, diagonalization and integration yield with

$$\lambda' = R^{-1}(-i\omega + \pi)R \tag{111}$$

an expression for $I(\omega, \tau)$:

$$I(\omega, \tau) = re\,\vec{W}\cdot\exp\{(i\omega + \pi)\tau\}R\exp\{\lambda'\tau\}(\lambda' - i\omega E)^{-1}\cdot R^{-1}\cdot\vec{1}. \tag{112}$$

Eq. (112) can be used for numerical calculations of the solid-echo spectrum and gives for the two frequency model discussed above expressions represented in [269]. The conclusions are the following:

The solid-echo spectrum $I(\omega, \tau)$ is the pure absorption spectrum $I(\omega)$ multiplied by a reduction factor depending on the pulse distance τ. This factor means a loss of intensity e.g. in the intermediate exchange region. Moreover, it contains a term proportional to ω^2 (ω being the distance of the mean frequency), leading to line shape changing compared with the FID spectrum. These differences are significant for mean transition rates and increase with growing pulse distances.

In this way the dynamic range in which investigations of correlation times and of types of motions are possible is extended to the limit T_2^*.

1.3.5.4 Ultraslow Motion

To extend the limit for the dynamic range from T_2^* for solid echoes up to the order of T_1, Spiess introduced the spin–alignment–echo technique (cf. Sect. 2.1.1) [270]. (For an agreement with the cited literature, we shall denote in this section the pulse distances with τ_1 and τ_2 instead of t_0 and t_1 in Table 2 and the times after the third pulse with t_2, after the spin–alignment echo with $t_2' = (t_2 - \tau_1)$. Assuming that $\tau_1 \ll T_2$, then the quadrupole frequencies between the first two pulses, as well as during the signal detection, hold constant expressed by

$$\omega_Q(\tau_2) = \omega_q = \delta(3\cos^2 \vartheta_q - 1), \tag{113}$$

$$\omega_Q(0) = \omega_{q'} = \delta(3\cos^2 \vartheta_{q'} - 1) = \omega_q + \Delta\omega_{q'}. \tag{114}$$

However, during the much longer period τ_2, characterized by spin–lattice relaxation with the time constant T_{1Q}, the quadrupole frequency can be changed by dynamic processes and strongly influences the spin–alignment echo. This echo can be expressed [270–272] by a single-particle correlation function

$$F(\tau_1, \tau_2) = \langle \sin[\omega_Q(0)\tau_1] \sin[\omega_Q(\tau_2)t_2] \rangle, \tag{115}$$

and the spin–alignment spectrum can be derived by Fourier transformation, beginning at the echo maximum

$$I(\omega, \tau_1, \tau_2) = \int_0^\infty F(t_2', \tau_1, \tau_2) \exp(-i\omega t_2') \, dt_2'. \tag{116}$$

$F(t_2', \tau_1, \tau_2)$ yields, by averaging all contributions from deuterons whose unique axis of the fgt forms an angle ϑ_q with the magnetic field $\vec{B}_0$ [271],

$$F(t_2', \tau_1, \tau_2) = \int [\langle Q_c^q \rangle \sin(\omega_q \tau_1) + \langle Q_s^q \rangle \cos(\omega_q \tau_1)] \sin[\omega_q(t_2' + \tau_1)] \, d(\cos \vartheta_q). \tag{117}$$

$\langle Q_c^q \rangle$ and $\langle Q_s^q \rangle$ are average phase factors depending on q and containing the information about the dynamic process. Therefore we need the conditional probability $P(\vartheta_{q'}|\vartheta_q,\tau_2)$ so that the corresponding angle would be ϑ_q at $t = \tau_2$ if it was $\vartheta_{q'}$ at $t = 0$:

$$\langle Q_c^q(\tau_1, \tau_2) \rangle = \int \cos(\Delta\omega_{q'} \tau_1) P(\vartheta_{q'}|\vartheta_q, \tau_2) d(\cos \vartheta_{q'}) \tag{118}$$

and an analogous expression for $\langle Q_s^q(\tau_1, \tau_2) \rangle$.

By varying τ_2 the decay rate of the spin–alignment echo and thus the correlation time can be determined, whereas variation of τ_1 enables one to distinguish different types of motion from the spin–alignment spectrum. There is, for example, a clear difference in the spectra for random small angular step reorientation and large angular jump motion with comparable correlation times. In connection with normal line shape measurements (FID) the spin–alignment echo permits the measurement of correlation times from 10^{-6} up to 10^2 s, i.e. over 8 orders of magnitude.

1.3.5.5 Distribution of Correlation Times

The influence of correlation time distributions on the quadrupole relaxation times can be treated like it was described in Sect. 1.3.3. These distributions, however, do also modify the shape of the solid-echo spectrum as well as its intensity [16, 273, 274].Whereas this influence can be neglected in the "rapid exchange" and in the "rigid solid limits", the changes are very important if the "intermediate exchange region" (10^{-6} s $\lesssim \tau_c \lesssim 10^{-5}$ s) is included in the correlation time distribution. For a heterogeneous distribution the resulting line shape represents a weighted superposition of line shapes for different correlation times which can simultaneously cover the slow, intermediate and fast exchange regions. By summarizing regions of correlation times inside which the spectra are practically constant, the superposition can be reduced to a weighted addition of less than 20 subspectra. These solid–echo spectra enable one to estimate the width of the correlation-time distribution and its symmetry. Moreover, a comparison of the line shape of partially and fully relaxed spectra can be used to distinguish between correlation time distributions which are homogeneous or heterogeneous in nature. Differences in these line shapes can be explained only by heterogeneous distributions because the correlation times are much smaller than the relaxation time T_1.

2 Experiment

In this chapter we want to mention briefly some requirements for NMR spectrometers, then to give a survey of the most important pulse sequences applied in the next chapters, and finally to discuss the particularities of some important methods of heterogeneous analysis in pulse NMR experiments. Although we want to speak shortly about some wide-line spectra in this book we will not discuss the continuous wave method here which measures the energy absorption during resonance crossing and represents the oldest NMR technique (cf. [97]). The requirements of the instrumentation for pulse NMR spectrometers are described in many papers (cf. e.g. [98]). For the *rf* transmitter the essential problems are very short *rf* pulses (sometimes, however, relatively long bursts) with very high power, different phases and well defined durations and distances. The sample probe must have one or more coils often for more than one frequency (e.g. also for decoupling and stabilization), the possibilities for wide range variable temperatures and for the magic angle sample spinning. Phase sensitive, or even quadrature detection, is necessary for a modern pulse spectrometer. The magnet with a good stabilization and a shim system ranges from air-coil over iron to superconducting systems for different applications. Very important is the data handling with various requirements for e.g. fast data acquisition, extended memory, computer controlled measurements and suitable signal processing (including data accumulation for the sensitivity enhancement) and representation. The various configurations depend on the applications.

2.1 Pulse Methods

The basic effect of the pulse methods is the rotation of the macroscopic magnetization around the $\vec{B}_1$-direction in the rotating frame about an angle $\alpha = \gamma B_1 t_w$ under the influence of an *rf*–pulse having the Larmor frequency $\omega_0 = \gamma B_0$, the amplitude B_1 and the duration t_w.

$\alpha = \pi/2$ denotes therefore a $\pi/2$ pulse etc. We suppose that for all experiments strong $\vec{B}_1$-fields are applied, i.e. all resonant nuclei are affected by the *rf*–pulses.

The different rotation axes in the rotating frame are caused by special *rf*–phases of the pulses. The magnetic field gradient lies in the direction of the static $\vec{B}_0$-field.

2.1.1 Nonselective Methods

Table 2 [99] gives a survey of some important pulse methods, which are, besides the WHH4-cycle [22], all of nonselective character. The solid-echo [100] and the

Table 2. Measuring methods in pulse-NMR

π	π-pulse	echo	spin-echo signal
$\left.\dfrac{\pi}{2}\right|_x$	$\dfrac{\pi}{2}$-pulse, applied along the X-axis in the rotating frame	$rf(t_1)$	long rf pulse of the duration t_1
t_1	distance between two pulses	$t_1(G)$	magnetic field gradient within two pulses separated by the distance t_1
FID	free induction decay	$[\ ]_n$	n-times repetition of the cycle

Quantity	Notation	Schematic representation	Determination of the measuring quantity
T_2	Free induction	$\dfrac{\pi}{2}$, FID	Time-constant of the FID
T_2	Hahn echo	$\dfrac{\pi}{2}, t_1, \pi, t_1$, echo	Time-constant of the echo maxima decay under t_1-variation
T_2	Solid-echo	$\left.\dfrac{\pi}{2}\right|_y, t_1, \left.\dfrac{\pi}{2}\right|_x, t_1$, echo	Ibid
T_2	Carr–Purcell–Meiboom–Gill	$\left.\dfrac{\pi}{2}\right|_y, [t_1, \pi|_x, t_1, \text{echo}]_n$	Time-constant of the echo maxima decay
T_1	Inversion recovery	$\pi, t_1, \dfrac{\pi}{2}$, FID	Time-constant of the FID maxima change under t_1-variation
T_1	Saturation recovery	$\dfrac{\pi}{2}, t_1, \dfrac{\pi}{2}$, FID	Ibid
T_1	Progressive saturation[1]	$\left[\dfrac{\pi}{2}, \text{FID}, t_1\right]_n$	Ibid
$T_{1\varrho}$	Spin-locking	$\left.\dfrac{\pi}{2}\right|_y, rf(t_1)|_x$, FID	Ibid

Table 2. (continued)

Quantity	Notation	Schematic representation	Determination of the measuring quantity
T_{2e}	MW4	$\left.\dfrac{\pi}{2}\right|_y, \left[t_1, \left.\dfrac{\pi}{2}\right|_x, t_1, \text{echo}\right]_n$	Time-constant of the echo maxima decay
T_{1D}	Jeener–Broekhaert	$\left.\dfrac{\pi}{2}\right|_y, t_0, \left.\dfrac{\pi}{4}\right|_x, t_1, \left.\dfrac{\pi}{4}\right|_x, \text{FID}$	Time-constant of the echo maxima decay under t_1-variation
Chemical shift parameters	WHH4	$\left.\dfrac{\pi}{2}\right|_y, \left[t_1, \left.\dfrac{\pi}{2}\right|_{-x}, 2t_1, \left.\dfrac{\pi}{2}\right|_x, t_1, \left.\dfrac{\pi}{2}\right|_y, 2t_1, \left.\dfrac{\pi}{2}\right|_{-y}\right]_n$	Fourier transformation of the envelope of the echoes between the $\left.\dfrac{\pi}{2}\right|_y$ and $\left.\dfrac{\pi}{2}\right|_{-y}$ -pulses
Analysis of heterogeneous systems	Goldman–Shen	$\dfrac{\pi}{2}, t_0, \dfrac{\pi}{2}, t_1, \dfrac{\pi}{2}, \text{FID}$	Cf. text in Sect. 2.2.2
D_s	Pulsed field gradient		Time-constant of the echo maxima decay under variation of the strength of the field gradient
	Hahn-echo	$\dfrac{\pi}{2}, t_1(G), \pi, t_1(G), \text{echo}$	
	Stimulated echo	$\dfrac{\pi}{2}, t_1(G), \dfrac{\pi}{2}, t_0, \dfrac{\pi}{2}, t_1(G), \text{echo}$	

[1] A special modification of the progressive saturation technique for solids which is favourable as to time consumption and accuracy is described in [103].

MW4-group [101, 102] (and the WHH4-cycle, of course) are specified for investigations of solids. The following measuring parameters are discussed:

T_1 the longitudinal relaxation time in the laboratory frame, the time constant to arrive at the thermal equilibrium;

T_2 the transversal relaxation time, describing the decay of the transversal magnetization to zero;

$T_{1\varrho}$ the longitudinal relaxation time in the rotating frame, practically a relaxation process in the *rf* field $\vec{B}_1$ that is essentially weaker than $\vec{B}_0$;

T_{2e} the effective relaxation time for special pulse groups (typically used for MW4), which represents the pulsed analogy to $T_{1\varrho}$;

T_{1D} the dipolar relaxation time [104], characterizing the relaxation process in the local magnetic field of the nuclear spins;

D_S the self-diffusion coefficient, describing the diffusion of the nuclear spins. The self-diffusion is generally non-coupled with a macroscopic transport of substance.

The determination of the second moment $\langle \Delta \omega^2 \rangle$ from the experimental data follows from Eqs. (7, 9).

For the investigation of quadrupole nuclei (e.g. deuterons) mainly the solid-echo and the Jeener–Broekhaert pulse sequences have been applied. The latter produces, for non-vanishing quadrupole interaction, the so-called spin–alignment echo, the maximum of which decays with the time constant T_{1Q}.

2.1.2 Selective Methods in Solids

Under high resolution conditions, discussed in Chapter 1, the Fourier transformation of the FID or of the half-echo gives the possibility of observing the change of separate lines and thus of determining the quantities under consideration selectively for the chemically nonequivalent nuclei. The WHH4-sequence as an example of multiple–pulse sequences for highly-resolved proton spectra is also defined in Table 2, representing the basic cycle for a lot of more complicated multiple–pulse sequences (cf. e.g. [11, 23]). Six different cross-polarization experiments are shown in Table 3. For all those the cross-polarization of the ^{13}C-nuclei via the protons in the preparation period characterized by the cross-relaxation time T_{CH} and the measurement of the ^{13}C–FID with dipolar proton decoupling are common. Sometimes for several reasons a simple $\pi/2$ (^{13}C) pulse instead of the cross-relaxation is applied. The phase-alternating *rf* field is coupled with a simultaneous change of the sign of data accumulation and leads to a suppression of artifacts [105, 106]. For a usual cross-polarization or T_1 experiment the time τ in Table 3 is equal to zero. The additional waiting period without any irradiation before data acquisition will be discussed in connection with the experiments in Sect. 3.4. The experiment proposed for the measurement of the cross-relaxation time also enables the determination of the proton spin–lattice relaxation time in the rotating frame $T_{1\varrho}$ under t_0-variation for longer t_0 values (cf. Sect. 1.2.6.1) selectively for several atomic groups whose ^{13}C-signals are chemically shifted.

Table 3. Cross-polarization experiments

(a) Common representation (abbreviations cf. Table 2)

	Preparation	Evolution	Detection
^{1}H rf-field	$\left.\dfrac{\pi}{2}\right\vert_x , rf\vert_{\pm y}$	See (b)	rf
^{13}C rf-field	$rf\vert_x$	See (b)	FID
Duration	t_0	t_1	t_2
Effect	Cross-polarization	Mixing	Dipolar decoupling

(b) Special evolutions:

Quantity	Evolution period	Determination of the measuring quantities from the Fourier-transformed FID(s)
Chemical shift parameters	$t_1 = \tau$ no irradiation (usually: $\tau = 0$)	Directly
T_2	No irradiation	Time-constant of the line intensities decay under t_1-variation
Chemical shift and dipolar splitting parameters	No irradiation or: ^{1}H: WHH4-sequence	After a second series of Fourier-transformations directly from the two-dimensional plot
$T_{1\varrho}$	^{13}C: rf-field	Time-constant of the line intensities decay under t_1-variation
T_1	^{13}C: $\left.\dfrac{\pi}{2}\right\vert_y , t_1 - \tau, \left.\dfrac{\pi}{2}\right\vert_y , \tau$ (usually: $\tau = 0$)	Time-constant of the line-intensities change under t_1-variation
T_{CH}	$t_1 = 0$	Time-constant of the line-intensities increasing under t_0-variation

2.2 Methods of Heterogeneous Analysis

Polymer system heterogeneity is determined by the polymer chain structure itself, by molecular-mass distribution, by phase structure, etc. Naturally this fact has an influence on the shape of NMR signals that often represent a sum of signals of various shapes and durations. The problem to analyse such complicated signals leads to the following:

(i) The signal shape with a certain heterogeneity type is to be established.
(ii) Signal components are to be attributed to corresponding structures.

(iii) Individual parameters of each component are to be determined. The solution
      of these problems is often not trivial and requires special methods.

2.2.1 Signal Decomposition

The problem of signal decomposition into components may be divided into two
cases:

(i) From the structure of the system it is *a priori* known that it consists of a limited
     number of discrete formations (partially-crystalline polymers, block-co-
     polymers, mechanical mixtures, etc.)
(ii) The nature of the multicomponent signal is not clear and may have different
     reasons that lead to discrete as well as to continuous distributions. (This case is
     usually due to heterogeneities on the molecular level.)

The problem of exactness and reliability of time signal decomposition into components was examined in detail in the sixties in a number of laboratories [107–109]
and was completely summarized in [110]. These works establish confidence
intervals which are applicable to the determination of decay parameters of various
magnetizations. Decomposition into discrete components is usually done by the
method of consecutive subtraction of the longest exponents or by the least square
method.

While, in the first case, these methods lead to quite reasonable results, in the
second case they may lead to erroneous conclusions. The method of the solution of
the second problem suggested in [111] is the most effective. This method permits

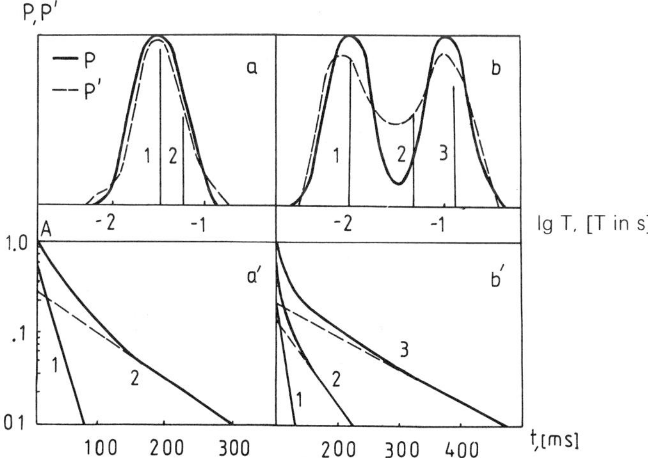

Fig. 9. Examples for the analyses of super Lorentzian curves (Distribution of relaxation times represented by one (a, a') and two (b, b') Gaussian curves) (cf. [112]). $P(T)$ initial (pseudoexperimental)
distribution, $A(t)$ calculated from $P(T)$ according to Eq. (119), $P'(T)$ calculated from $A(t)$ by inversion of
Eq. (119), *1, 2, 3* relaxation times and their intensities obtained by traditional discrete analysis

the analysis of curves such as

$$A(t) = \int_0^\infty P(T)\exp\left(-\frac{t}{T}\right)dT \tag{119}$$

without knowledge of the type of distribution. This method was examined in detail and improved in [112]. There were analysed pseudoexperimental curves such as (119) having various types of distributions $P(T)$. The comparison with the deconvolution of decays into discrete components was done graphically. Figure 9 shows the most characteristic results. It can be seen that the method correctly represents the form of the distribution function, while the components obtained by graphic analysis are disposed quite arbitrarily in given spectra limits.

It was supposed that this method is useful in the case of a signal that is determined by the sum of several discrete distributions. Analogue problems exist also for the decomposition of wide-line NMR spectra. Since these questions do not play such an important role in this book, we refer here only to the basic papers of Bergmann [113, 114]. However, the question of the line shape caused by the anisotropy of the chemical shift and influenced primarily by order and orientation parameters is discussed in greater detail in Chapters 1 and 3.

2.2.2 Two-Dimensional Analysis in Time Domain

This section will discuss the methodical approach enabling us to solve the question which structural phases of the sample the components of the signal may be attributed to. These methods are based on the comparison between the decay shapes of longitudinal and transversal magnetization. The fact is that attribution of the components of transversal magnetization does not usually cause difficulties since in the overwhelming majority of cases the component with a longer relaxation time corresponds to the most mobile phase. For the longitudinal magnetization the attribution becomes ambiguous because the shape of the temperature dependence of spin–lattice relaxation looks like a "V". In many cases it is quite enough to establish the correspondence between the components of longitudinal and transversal magnetizations to restore the unambiguity of component attribution for the longitudinal relaxation.

The essence of the method is as follows: the signal from one or another sample phase can be observed only when the time passed from the beginning of the spin system excitement till the observation moment is less than the time of spin–lattice relaxation of this phase. Hence, the examination of the change of transversal magnetization decay shape being carried out during the measuring process of longitudinal magnetization can show that each component of the longitudinal magnetization corresponds to one component of the transversal decay [115, 121].

Figure 10a shows the scheme of such an experiment. It should be noted that besides the phase attribution this method reveals if either spin–lattice relaxation or spin–dynamics processes determine a given component. If the component is determined by a spin–dynamics process, the transversal magnetization component corresponding to this phase does not vanish completely in the time scale

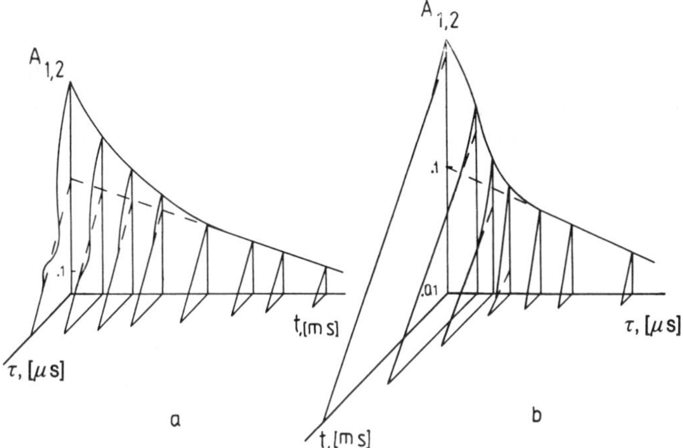

Fig. 10. Schemes of the experiments for the analysis of multicomponental magnetization decays in the time domain. Axis t: longitudinal, axis τ: transversal magnetization

considered. Another version of this method consisting of the examination of the decay shape of the longitudinal magnetization changing during the transversal magnetization damping is suggested in [116]. This method is shown in Fig. 10b and may be useful to determine the spin–lattice relaxation times when for some reasons longitudinal relaxation decomposition into components goes beyond the confidence interval.

The so-called Goldman–Shen [117] experiment is another version of the two-dimensional analysis of decays. This method can be applied to the analysis of systems containing two-component transversal magnetization and one-component longitudinal magnetization provided that $T_1 \gg T_{2a} \gg T_{2b}$. The experiment scheme is shown in Table 2. The essence of the method is that transversal magnetization being transferred to the XY plane by a 90°-pulse applied to spin system is damped freely on this plane. Then after a time interval t_0 that is longer than T_{2b}, but shorter than T_{2a}, a 90°-pulse is applied again and returns phase "a" magnetization to the Z-axis. In a time t_1 longer than T_{2a} and less than T_1 a 90°-pulse is applied once again, and transverse magnetization decay is observed. If the· process of spin diffusion between the two phases is slow, the signal observed after the last applied pulse must repeat the shape of the slow component, while the component with the short T_{2b} must appear additionally in the opposite case. This experiment may give information on the nature of the nonexponentiality of magnetization decays as well as structure information on the sizes of these heterogeneities. The most successful application of this method was the analysis of heterogeneous polymer systems in [118–120]. In a similar manner also proper modifications of cross-polarization experiments can be used to distinguish parts of the signal which characterize regions in the polymer molecule with different mobility or molecular order. How one can use either the carbon relaxation times T_1 or T_2 or the cross relaxation time T_{CH} for such discriminations is shown in Sect. 3.4.1.

2.2.3 Two-Dimensional Spectroscopy

Two-dimensional (2D) spectroscopy is described in the fundamental work of Ernst and coworkers [122] as a representation of a data set of two frequency variables, independent of the kind of measurement. Often the measuring variables are two times t_1, t_2 which are connected with the frequencies v_1, v_2 by a twofold Fourier transformation

$$g(v_1, v_2) = \int_0^\infty dt_1 \exp(-2\pi i v_1 t_1) \int_0^\infty dt_2 \exp(-2\pi i v_2 t_2) M(t_1, t_2), \tag{120}$$

where $M(t_1, t_2)$ is the measured function of two time variables and $g(v_1, v_2)$ is the derived function of two frequency variables. The two-dimensional experiment consists generally of a preparatory period, then an evolution period with the duration t_1 under the influence of the Hamiltonian $\mathscr{H}_1$ and at last a detection period with the running time t_2 during which the Hamiltonian $\mathscr{H}_2$ is effective. That means many repetitions of the experiment for different evolution times t_1 are needed to get a two-dimensional plot of spectral data apart from the data accumulation for the enhancement of the sensitivity. Hence, the total time for a 2D experiment is very long. That is the price, however, for a drastic growth of information on the molecular structure. This basic idea, initially applied to NMR spectroscopy in liquids, provides a large number of experiments in dependence on the formation of the three time periods, e.g. using different *rf*-pulse sequences.

We shall consider a modified cross-polarization experiment proposed by Waugh and coworkers [123, 124], cf. also Table 3, which is very similar to that performed for the measurement of the ^{13}C spin–lattice relaxation time in the rotating frame $T_{1\varrho} - {}^{13}$C. The preparation consists of the usual procedure of cross-polarizing the carbons using a spin-locking proton experiment connected with ^{13}C-irradiation under Hartmann–Hahn conditions. During the evolution period t_2, *rf*-fields neither for protons nor for carbons are applied, that means the dipolar interaction as well as the chemical shielding are effective. Sometimes a WHH4 four-pulse cycle to the protons was applied simultaneously to suppress the dipolar proton–proton couplings. Finally the data acquisition takes place under dipolar decoupling of the protons, i.e. only chemical shift is working in this period.

For the Hamiltonians at the time periods t_1 and t_2 we get

$$\mathscr{H}_1 = \mathscr{H}_{cs} + \mathscr{H}_D^{IS} \tag{121}$$

and

$$\mathscr{H}_2 = \mathscr{H}_{cs}, \tag{122}$$

with Hamiltonians for the chemical shift $\mathscr{H}_{cs}$ and for the dipolar proton–carbon interaction $\mathscr{H}_D^{IS}$, which are given by

$$\mathscr{H}_{cs} = \sum_n \Delta_n S_{zn} \tag{123}$$

and

$$\mathscr{H}_D^{IS} = \sum_n S_{zn} \sum_k (-2b_{nk}) I_{Zk} \tag{124}$$

with Δ_n being the chemical shifts of the ^{13}C-spins, and b_{nk} is given by Eq. (5) containing the structural information. Denoting the latter sum over K (substituting I_{Zk} by the corresponding magnetic quantum number) by Ω_n, one gets for the measured signal

$$M(t_1, t_2) = \sum_n \exp\{i[\Delta_n t_2 + (\Delta_n + \Omega_n)t_1]\} \tag{125}$$

and after the two-dimensional Fourier-analysis according to Eq. (120)

$$g(v_1, v_2) = \sum_n \delta(v_1 - \Delta_n - \Omega_n)\delta(v_2 - \Delta_n). \tag{126}$$

The two-dimensional frequency plot therefore shows pure chemical shifts in the v_2-axis but in v_1-direction the dipolar broadenings Ω_n with respect to the chemical shift Δ_n, where the dipolar signals are arranged with respect to the line $v_1 = \Delta_n(= v_2)$ as it is depicted schematically in Fig. 11. Whilst in the high resolution experiments the dipolar interaction is usually lost, with the so-called "separated local field method" [123, 124] one can measure the dipolar splitting separately for each chemical shift and therefore obtain structural information for each chemical nonequivalent ^{13}C nucleus regarding its joining vectors with the neighboring protons. Summing up the spectra vertically or horizontally one obtains the pure chemical shift spectrum and a wide-line spectrum with dipolar broadenings, respectively.

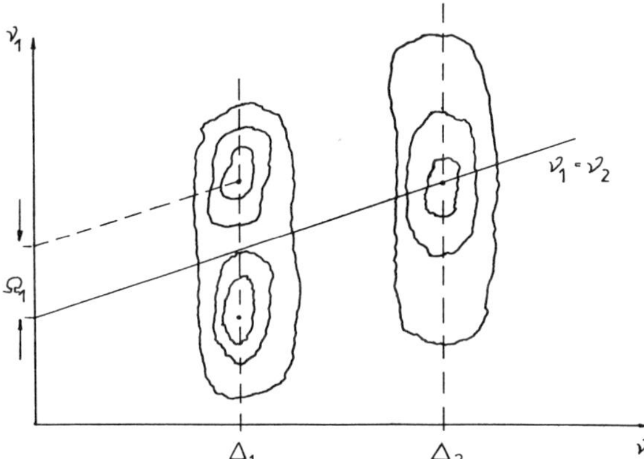

Fig. 11. Contour plot of the function $f(v_1, v_2)$ with a doublet at Δ_1 and a singlet at Δ_2 (concerning v_1, v_2, Ω_1 cf. text)

In a similar manner it is also possible to regain the interesting information on the anisotropy by MAS-technique for resolution enhancement. This 2D experiment is described by Lippmaa and coworkers [125].

Finally it should be remarked that the experimental requirements are comparable with those for the measurement of selective relaxation times.

Recently a new two-dimensional MAS–NMR experiment has been introduced [281, 282] in which a waiting period t_1 was inserted just before the start of a usual cross-polarization pulse sequence relative to a fixed point in the rotor period. In the case of isotropically disordered samples as well as for unsynchronized experiments over a sufficient number of transients the signal would be unmodulated during t_1. For partially ordered systems, however, with a preferential axis of orientation not parallel to the rotor axis the measured sideband intensities depend on the instant angle which describes the rotation of the preferential axis system about the rotor axis. Therefore, this method allows one to determine the angle of such a preferential axis (director) with respect to the rotor axis, the orientation of the chemical-shielding tensor relative to the preferential axis and the degree of order in the sample. The disorder can be expressed by a characteristic parameter of a spherical Gaussian distribution describing the root-mean-square angular deviation of the order axis or by the distribution of two angles correlating the director with regard to the molecular frame. The experiments can be optimized by suitably adjusting the two parameters, i.e. the spinning frequency and the angle between the rotor and the director axis.

Whereas the correlation of two tensorial interactions (e.g. anisotropic shielding and dipolar coupling) is considered in the two-dimensional techniques mentioned above, a correlation of one interaction with itself can yield information on dynamic exchange processes. If the molecular motion is characterized by jumps between different positions, a corresponding two-dimensional deuteron experiment [279] with a pulse sequence like that for the spin–alignment echo (with variation of the first pulse interval) shows, besides a deuteron powder pattern, some ridges in the form of two ellipses centered at $\omega_1 = \omega_2 = \pm\frac{1}{2}\delta$, with the semi-axes $(3/\sqrt{2})\delta\cos\theta$ and $(3/\sqrt{2})\delta\sin\theta$, respectively. Therefore, the angle θ describing the rotation of a unique axis (e.g. a symmetry axis of a fast rotating methyl group) during the mixing time between the second and third pulse can be read out directly from the 2D plot. Its tangent equals the ratio between the two semi-axes. Thus, these jump angles can be determined with high accuracy. Moreover, since different types of molecular motion lead to different 2D exchange spectra the diffusive motion can already be discriminated from discrete jump motion using the qualitative impression of the 2D plot.

3 NMR Signal Shape and Polymer Structure

3.1 Transverse Magnetization of Protons

Investigation of the shape of the transversal magnetization decay (TMD) or of proton broad lines in polymer systems has advanced due to the solution of two general questions that appeared in the 1970s. They are, firstly, the so-called "solid effect" found in polymers at temperatures higher than the glass-transition and melting temperatures (T_g and T_m) [59, 126–132, 319], and secondly, the working out of multicomponent analysis methods of partially crystalline polymers [114, 133–136].

The present section generalizes the TMD shape investigations in order to establish relations between the signal shape and peculiarities of the molecular and phase polymer structure. The correct signal decomposition into components as well as the proper attribution of the latter to corresponding structure formations are the most important problems of spectra analysis, since the polymers frequently have a heterogeneous structure. Here we will discuss the effects of heterogeneity of a different nature that is due to the molecular structure of the sample as well as to the phase structure. We will analyze TMD shape variations in dependence on the external influences (temperature, swelling, stretching, orientation) and on internal structure of macromolecules and crystals.

3.1.1 Experiments

The typical TMD shape dependences on temperature for amorphous polymers (rubber and thermoplastics) are given in Fig. 12, while Fig. 13 represents the same dependences for crystalline polymers (LPE and PVF). TDM dependences on the crystallinity for PETP and on the orientation in the magnetic field for oriented LPE are shown in Fig. 14 (cf. Table 4 also). TMD dependences on molecular weight for PE and PEO melts are demonstrated in Fig. 15. TMD shape dependence on relative lengthening due to monoaxial stretching at room temperature obtained for crosslinked polyisoprene rubber is given in Fig. 16.

All the main particularities revealed in the whole series of experiments are qualitatively shown in Fig. 17. Here are the most general of them:

(i) The TMD shape varies within Gaussian, Lorentzian (exponential) and super-Lorentzian (sum of exponentials) forms for amorphous polymers.

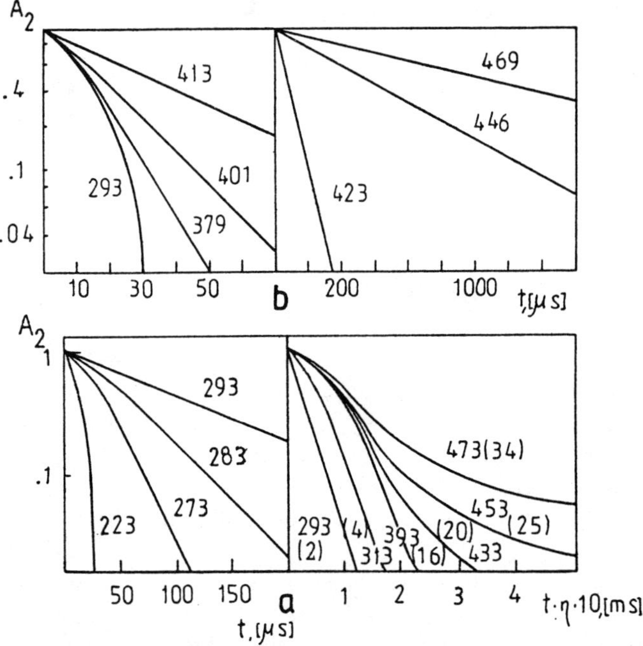

Fig. 12. Transversal magnetization decay in (**a**) PIB [59] and (**b**) atactic PS [60] for different temperatures. Parameters on the curves: measuring temperature in K, in parentheses: scaling factor η

(ii) The TMD shape does not depend on the chain structure at $T \lesssim T_g + 50\,\text{K}$ but does so at higher temperatures. The Gauss-like component appears in rubber at $T > T_g + 100\,\text{K}$ while in thermoplastics it is absent.

(iii) At low temperatures FID (TMD) of crystalline polymers is described by an oscillating function.

(iv) The oscillating component depends on the orientation in the field $\vec{B}_0$ and vanishes after transition to the amorphous state (amorphization or melting take place).

Table 4. FID parameters from the curves given in Figs. 13 and 14 according to Eqs. (127, 128). Second moments in $10^{10}\,\text{s}^2$, T_2 in μs

Polymers	T [K]	$b^2/3\,(b^2)$ cr	$\langle \Delta\omega^2 \rangle$ cr	int	am	P_{cr}	P_{int}	P_{am}
LPE (0°)	173	2.13	2.3	1.34	1.62	0.60	0.28	0.12
LPE (45°)	173	0.99	1.55	1.82	1.59	0.63	0.26	0.11
LPE (90°)	173	1.62	2.5	2.12	1.69	0.70	0.18	0.12
LPE (isotr.)	173	1.57	1.93	1.78	1.64	0.70	0.12	0.18
PVF (isotr.)	173	(0.85)	1.4	1.17		0.5	0.5	
PVF (isotr.)	299	—	1.12		0.15	0.73		0.27
PVF (isotr.)	373	—	0.87	(53)	(200)	0.45	0.28	0.27
PETP (isotr.)	163	(0.44)	0.74	0.72		0.55	0.45	

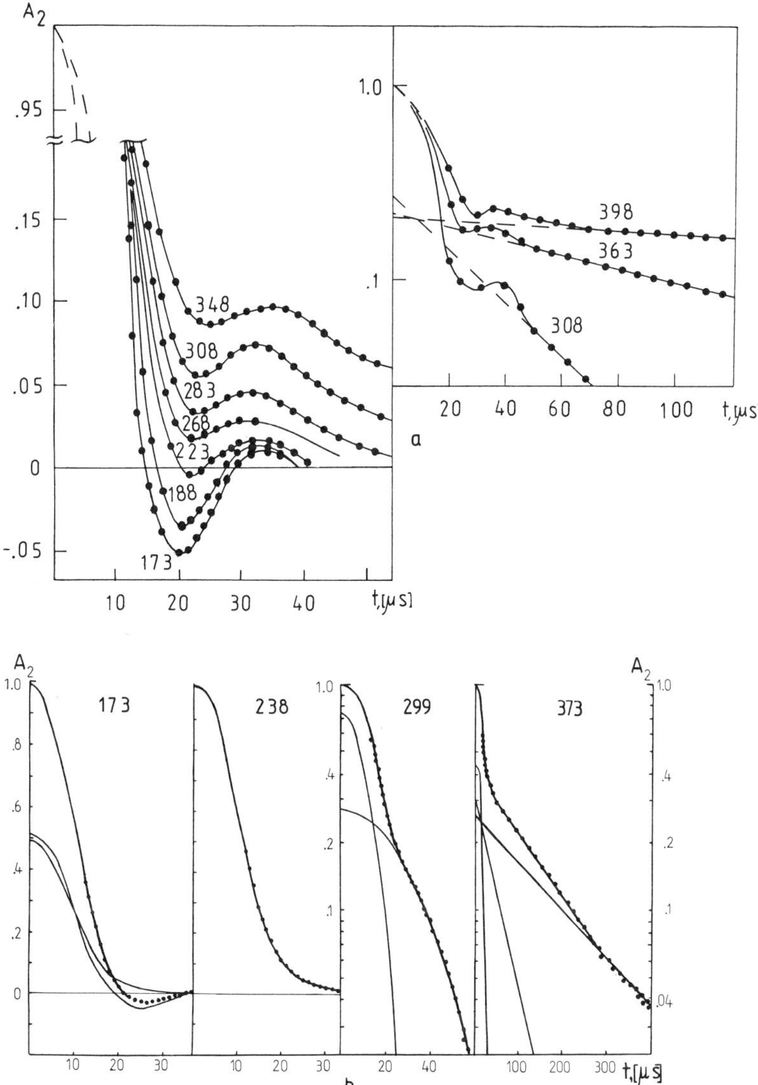

Fig. 13. Free induction decay in (a) LPE [134, 135] and (b) PVF [137] for different temperatures (Denotations cf. Fig. 12). Solid lines: FID according to Eqs. (127, 128)

(v) One can easily see that TMD of crystalline polymers as well as of amorphous polymers splits into several components at high temperatures.

Let us try to clear up the reasons that affect the TMD shape. According to the theory the Gaussian line shape is characteristic for solids, and the Lorentzian line shape for liquids in the substances with a high concentration of paramagnetic centres. The Lorentzian line shape is observed only for magnetic-diluted systems in solids, though the substances consisting as a whole of paramagnetic centres can

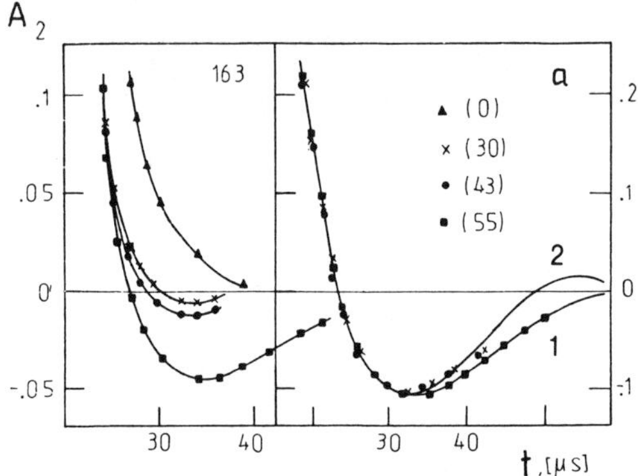

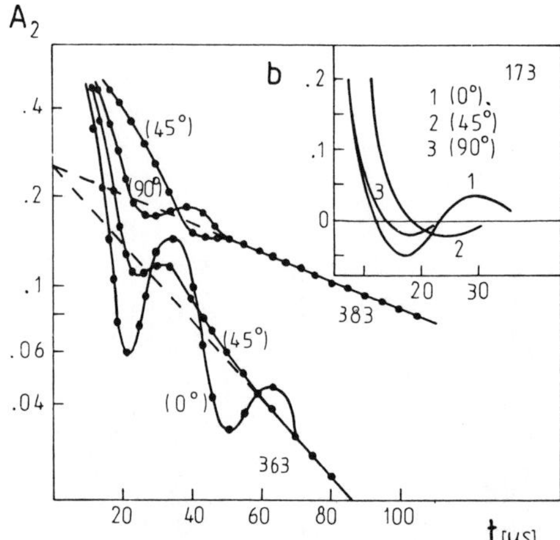

Fig. 14. Free induction decay [60] (**a**) in PETP and (**b**) in oriented LPE. (**a**): temperature 163 K, left: FID of the common sample, right: FID of the crystalline component only. Parameter on the curves in parentheses: degree of crystallinity, (**b**): Parameter on the curves: measuring temperature in K, in parentheses: angle of the orientation with regard to the magnetic field. Solid lines: FID according to Eqs. (127, 128) using parameters from Table 4 for PE and $a^2 = 3 \times 10^9 \text{ s}^{-2}$, $b^2 = 4.4 \times 10^9 \text{ s}^{-2}$ (1) and $a^2 = 1.2 \times 10^9 \text{ s}^{-2}$, $b^2/3 = 6 \cdot 10^9 \text{ s}^{-2}$ (2) for PETP.

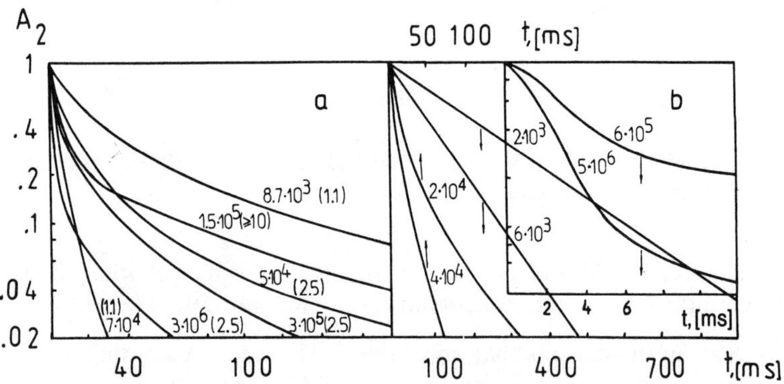

Fig. 15. Transversal magnetization decay for (a) LPE [60, 42] and (b) PEO [59] for different molecular masses. Parameter on the curves: molecular masses, in parentheses: ratio M_w/M_n. Arrows show the corresponding time-axes

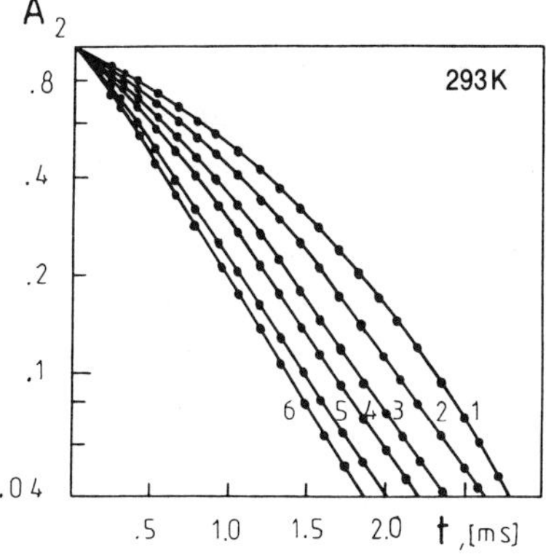

Fig. 16. Transversal magnetization decay in cross-linked PIP rubber for different stretching factors [138]. Parameters on the curves: stretching factor

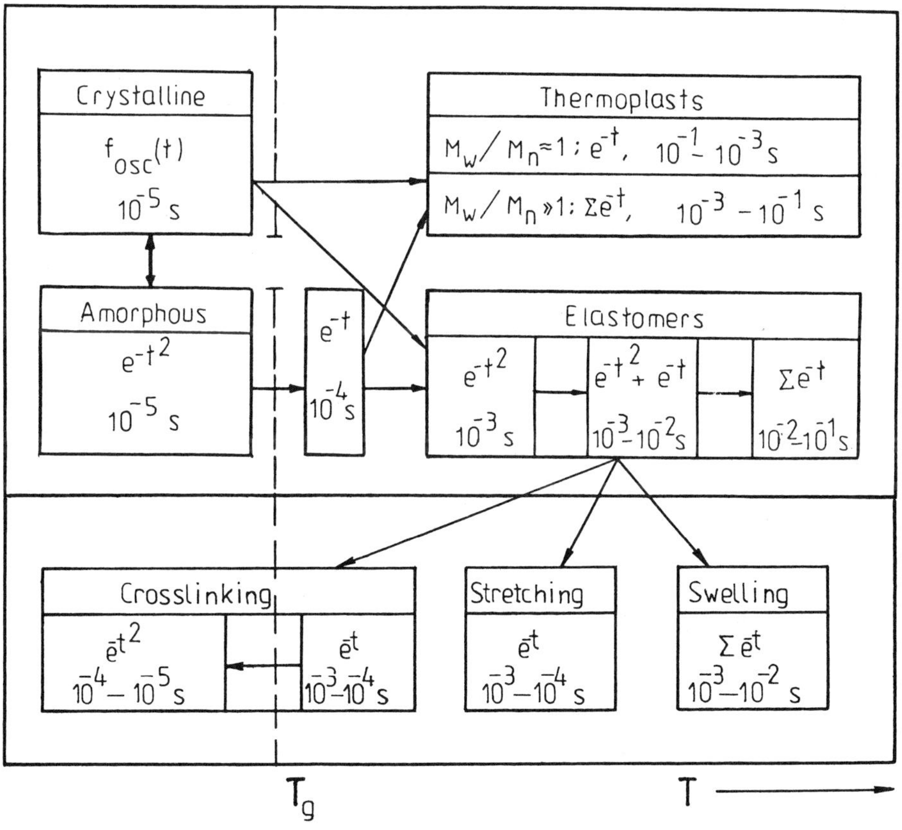

Fig. 17. Typical shape of the transversal magnetization decay and its variations in bulk polymers

behave as quasidiluted [55]. Hence, there are two possibilities to explain TMD shape changing from Gaussian to exponential shape. An appearance and development of rapid molecular motion is one of these peculiarities while another is due to a change of such spin system parameters that leads to a quasidiluted solid spin system.

The criterion of distinction between these two reasons is the presence or absence of a solid-echo signal, which is caused only by dipole–dipolar interaction not averaged by molecular motion. This case shows that the envelope of solid-echo amplitudes is damped much slower than is the usual transversal magnetization. In this case of rapid isotropic motion a solid-echo (SE) does not arise after the second $90°$ pulse. The results of such an experiment for various polymer systems are given in Fig. 18. Thus the decay of TMD in solid PS and PE, as demonstrated in Fig. 18, is several times faster than for the envelope of SE. The difference between damping rates decreases at higher temperatures and vanishes completely at temperatures $\approx T_g + 50\,\mathrm{K}$, and the TMD shape becomes Lorentzian. Similar behaviour is observed over this temperature range for all polymers. At higher temperatures SE does not appear in thermoplastics, while it is regenerated in rubber at $T > T_g + 100\,\mathrm{K}$ (the TMD shape becomes Gauss-like); this reveals that non-averaged dipole–dipolar interactions are present in the system. The solid-echo appears also in cross-linked polymers with the TMD shape transferred from a Gaussian to an exponential form under swelling or stretching.

A similar experiment carried out at high temperatures showed that the envelope of SE was not very different from the FID in non-crystalline regions, if its shape preserved an exponential look. That experiment enables us to conclude that the transition of the TMD from a Gaussian to an exponential shape with tempera-

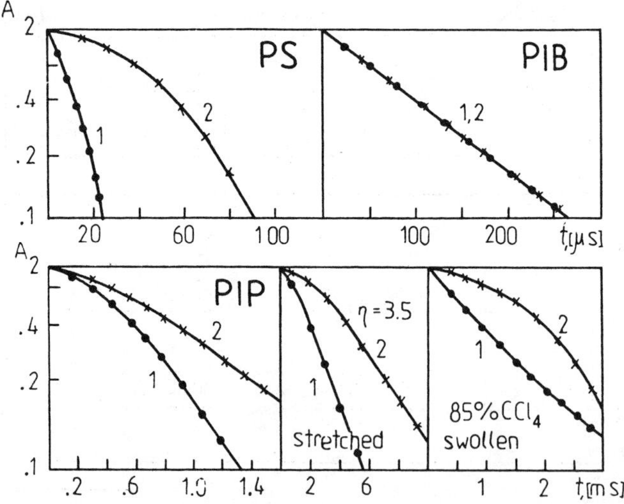

Fig. 18. Transversal magnetization decay (*1*) and envelope of the solid-echo (*2*) in some typical situations in amorphous polymers [60] at 293 K.

ture increase is generally due to molecular motion for all polymers, while a similar transition of the TMD shape observed under cross-linking, swelling or stretching in rubber is due to variations of the interactions in the internal spin system resulting in the conditions of quasidiluted spin systems. This indicates that the signal from such a sample is a sum of signals with Gaussian or Gauss-like shape. The study of NMR spectra in crosslinked PS [129] obtained by sample rotation about the magic angle leads to the conclusion that the broadening of NMR super-Lorentzian line shape has a static character.

3.1.2 Analysis of the Experiments in Semicrystalline Polymers

Let us analyse the TMD shape in crystalline polymers with the principal aim of decomposing the signal into components correctly and attributing the latters to certain structures. Recently, two approaches to a solution of this problem were demonstrated in the literature. One analyzes broad line spectra, the other pulse NMR data.

The first method is based on the concept of the phase structure according to which the latter is identified either in the presence or in the absence of molecules with a certain type of motion. This causes strong temperature dependence of the intensities of the signal components that in their turn produce the notion of dynamic crystallinity [139, 140] in contrast to the static one obtained from structure methods.

The logic of some other approaches, developed in a series of pulse NMR publications [134, 135] will be demonstrated later by an example of FID shape analysis in linear PE.

Figure 13a shows that the FID shape in PE at high temperatures can be represented by a sum of three components: one oscillating and two exponential. The intensities of these components determined by graphic analysis were compared with the intensities of the three components of a wide-angle X-ray scatterring (WAXS) spectrum obtained for the same sample.

The intensity of the oscillating FID component appeared to be equal to the corresponding of the crystal WAXS spectrum component, and the intensity of the weakly damped exponential term coincides with that of the amorphous WAXS spectrum component. One can conclude that FID components are connected with the crystalline, intermediate and amorphous phases of PE. Furthermore one can logically assume that the correspondence of FID components and structure phases is retained over the whole range of temperatures lower than the melting temperature, with FID shape of each component being determined by the peculiarities of the interaction of the internal spin system of each phase under certain conditions. Variation of the level of the molecular motion in each phase, as it follows from the above, must not vary the number of given protons in these phases, though in principle relative intensities measured in the experiment can vary, which is due to the coincidence of signal shapes or to the similarity of their time parameters (see e.g. the evolution of the FIDs in PVF given in Fig. 13b).

The next stage of the analysis is to find the FID shape for each component and to determine the signal parameters. Since FID shape calculations in solids cannot be carried out exactly, we shall use empiric functions to describe the experiment quantitatively. The criterion of the correct selection of the function is the coincidence of moments of these functions with the values calculated by means of the Van-Vleck formulas [9].

The selection of these functions for each signal component made in [133–135] for polyethylene, in [38] for poly(ethylene terephthalate) and in [137] for poly(vinylidene fluoride) shows the possibility of describing the TMD shape in partially crystalline polymers in the whole temperature interval by the expression:

$$A_2(t) = \sum_i P_{2i} A_{2i}(t) \qquad (i = \mathrm{cr, int, am}), \tag{127}$$

where $P_{2\mathrm{cr, int, am}}$ and $A_{2\mathrm{cr, int, am}}$ are the intensities and the shapes of the TMD components of the crystalline, intermediate and amorphous phases.

For the crystalline line shape yields (as to the moments cf. Table 1):

$$A_{2\mathrm{cr}}(t) = \begin{cases} \exp\left(-\dfrac{a^2 t^2}{2}\right) \dfrac{\sin bt}{bt} & (128\mathrm{a}) \\[2em] \exp\left(-\dfrac{a^2 t^2}{2}\right) \cos bt & (128\mathrm{b}) \\[2em] \exp\left(-\dfrac{a^2 t^2}{2}\right) & (128\mathrm{c}) \end{cases}$$

Eqs. (128a, b) are characteristic for low temperatures while Eq. (128c) is valid near the melting temperature. The line shape for the intermediate and amorphous components is given by

$$A_{2i}(t) = \exp\left[-\langle \Delta\omega^2 \rangle_i \tau_i^2 f(t/\tau_i)\right] \qquad (i = \mathrm{int, am}) \tag{129}$$

with $f(t/\tau_i)$ from Eq. (70). τ_i is the correlation time of the segmental motion. Equation (129) at temperatures lower than T_g ($\tau_i \gg t$) becomes $\exp(-\langle \Delta\omega^2 \rangle_i t^2/2)$ and at temperatures higher than T_g ($\tau_i \ll t$) $\exp(-\langle \Delta\omega^2 \rangle \tau t)$.

The analysis of the experimental curves is made using Eq. (127) by the least square method and results in the determination of seven adjustable parameters characterizing the spin system of all three polymer phases ($P_{2\mathrm{cr, int, am}}$, $\langle \Delta\omega^2 \rangle_{\mathrm{cr, int, am}}$, b^2).

The study of temperature and orientation dependences obtained by the adjustment of the parameters of each FID component of the spin system and the comparison of these dependences with the theoretical ones or with the results of other methods leads to the following conclusions:

(i) It is established that the intensities of all three components in LPE and PVF within the experimental error limits do not depend on the temperature (cf. Fig. 13) and always correspond to the number of polymer chains in each phase.

LPE, in which the effect has been examined in detail, reveals the coincidence of the intensities of the FID components with those of the WAXS spectra measured at high temperatures (the temperature dependence of the intensities of the crystalline and intermediate phases is obtained in WAXS).

(ii) The shape of the crystalline FID component is described by Eq. (128a) in PE and PEO and by Eq. (128b) in PETP and PVF. The second moments are in agreement with those calculated by means of the Van–Vleck formulae for all polymers. The quantities $b^2/3$ determined for PE and PEO coincide with the intramolecular contribution to the second moment, and the quantities b^2 determined for PVF are close to the CH_2 doublet splitting.

(iii) The TMD shape is always Gaussian for both noncrystalline components at low temperatures ($T < T_g$). Here the second moments in oriented LPE reveal orientation dependences for the intermediate phase while they do not do so for the amorphous one (cf. Table 4).

(iv) The TMD shape of noncrystalline components at high temperatures ($T > T_g$) is nearly exponential. Here orientation dependences of $T_{2,\,int,\,am}$ do not exist though WAXS spectra show orientation dependence of the intermediate phase under these conditions.

3.1.3 Nature of the Approximating Functions in the Solid State

It is natural to expect that the FID is described by Eq. (128b) in all cases since the polymers we regard are mainly based on CH_2 groups, the interactions of which determine the FID shape.

However, such a shape is only observed in PETP and PVF crystals and nowhere else. The FID shape is Gaussian in all amorphous polymers while it is described by Eq. (128a) in PE and PEO crystals.

Equation (128a) represents a superposition of the Fourier-transforms of a normal and a rectangular distribution function. It can properly describe the FID shape in ionic crystals. As presented in [141], Eq. (128a) has a simple physical sense:

The local field which affects the reference spin in the crystal consists of two parts, one of which corresponds to the influence of the neighbouring and the other of the more distant spins.

The cell of the nearest spins (here called reference cell) is formed according to the strength of the interaction of the reference spin with the others. Calculations made for crystalline CaF_2 showed that the parameters a^2 and $b^2/3$ of Eq. (128a) are connected with the lattice sum in such a way that $b^2/3$ is equal to the contribution of the spins of the reference cell to the second moment. In PE one can clearly see that the equality of $b^2/3$ and of the intramolecular contribution to the second moment is not accidental. Therefore we can conclude that the reference cell in a polymer crystal is that section of a macromolecule in which dipolar interactions along the chain are correlated. This correlation is caused by the dynamic term of the dipolar interaction.

From these presentations one can easily understand the reasons for the differences in the FID shape determined by spin system structure of a macromolecule chain. In the sequence PE-PEO-PETP the chain structure varies in such a way that the distance between the glycol groups increases. In the PEO chain the influence of the oxygen atom on this distance is not too great, because of the spiral conformation of the chain in the crystal. However in PETP the glycol groups are separated by great distances that are determined not only by ether groups but also by weakly interacting benzene rings, and hence, in PETP these groups can be considered as being isolated.

The FID shape in PVF where there is no spatial isolation is determined by the fact that the dipolar interaction of protons and fluorines has no dynamic term and, as a result, the reference cell is formed by a pair of protons.

The Gaussian shape of FID observed in these polymers after the transition from the crystalline to the amorphous state shows that the correlation of the nearest spins is disturbed completely in the latter and that the concept of the reference cell no longer applies. The second moments for a polymer in the amorphous state differ slightly from those for the same polymer in the crystalline state. In the case of PETP the second moment in the amorphous state is even larger, approximately 20%, than in the crystal state although the macroscopic density of a crystalline polymer is higher than of an amorphous one. In our view, such behaviour of the FID shape and of the second moments can be explained only on the assumption of irregularities in the distances between the protons. These irregularities may be of intramolecular as well as of intermolecular origin. The first may occur because of irregularly alternating gauche- and trans-positions of the CH_2 groups in the amorphous state. The second may be connected with density fluctuations that shorten the distances of CH_2 groups as compared to their distances in crystalline materials. Both reasons lead to the disappearance of the reference cell of correlative spins, which belongs to the same macromolecule and is connected with the strictly determined configurations of the atomic groups.

To evaluate the effects connected with the irregularities of the intramolecular origin quantitatively we numerically calculated the line shape for a system of four spins, which corresponds to the configuration of the glycol group which is found in *trans-* and *gauche*-conformations in PETP. In both cases the obtained spectra that are averaged for all directions represent slightly differing doublets with well resolved peaks. The second moments of these two spin configurations are also very close. It can be concluded that an appearance of a Gaussian line shape in the amorphous state is determined first of all by intermolecular irregularities that are due to density fluctuations.

3.1.4 Approximating Functions and Semiquantitative Description of Experiments in Polymer Liquids

TMD shape evolution in amorphous polymers and in noncrystalline regions of semi-crystalline polymers at temperatures higher than T_g or T_m can be described by means of a simple theory [59] based on the following assumptions:

(i) The spin system of the amorphous sample at temperatures higher than T_g is heterogeneous and consists of spin systems of macromolecules or of their large fragments. Each of these macromolecules can be characterized by an averaged local field.

(ii) The local field of each spin system consists of two independent components, fast and slow; their fluctuations represent a Gauss–Markovian process and are characterized by the correlation times τ_f and τ_s.

(iii) Fast fluctuations of the local field are determined by segmental anisotropic motions of the chain sections while slow fluctuations are determined by "large-scale" motions that average the remaining part of interaction. The formation of the spin system of the macroscopic sample depends on the mechanism of the mutual interaction of these spin systems and this determines the magnetization decay.

The following two limiting cases will be discussed:

(i) Spin systems of macromolecules form a common spin system owing to strong intermolecular interaction.

(ii) Spin systems of macromolecules are not connected with one another, i.e. the intermolecular interaction is many times smaller than the intramolecular one.

In the first case the TMD expression which characterizes the sample as a whole according to Eq. (63) is

$$A_2(t) = \exp\{ - \langle \Delta\omega^2 \rangle [q^2 \tau_f^2 f(t/\tau_f) + (1 - q^2)\tau_s^2 f(t/\tau_s)]\}, \tag{130}$$

where $\langle \Delta\omega^2 \rangle$, τ_f, τ_s, and q^2 are the weight averaged values of the parameters which characterize the common spin system of the whole sample. According to Eq. (130) the TMD shape can vary only within Gaussian and exponential.

In the second case the sample spin system is composed of N independent systems, and the common signal consists of the sum of independent $A_2(t)$ signals of each spin system:

$$A_2(t) = \sum_i P_{2i} A_{2i}(t), \tag{131}$$

where P_{2i} is the weight of i-th spin system and $A_{2i}(t)$ is determined by an expression of the form of Eq. (130). Hence, TMD may be of any shape from Gaussian to super-Lorentzian according to the type of P_{2i}-distribution.

At temperatures close to T_g the TMD of all amorphous systems is completely determined by the expression

$$A_2(t) = \exp[- \langle \Delta\omega^2 \rangle \tau_f^2 f(t/\tau_f)] \tag{132}$$

that is obtained from Eq. (130) if $\tau_f^2 \langle \Delta\omega^2 \rangle q^2 \approx 1$, $\tau_s \gg \tau_f$.

At higher temperatures ($T \gtrsim T_g + 100\,\mathrm{K}$) the conditions $\tau_f^2 \langle \Delta\omega^2 \rangle q^2 \ll 1$ and $\tau_s^2 \langle \Delta\omega^2 \rangle (1 - q^2) \approx 1$ are valid and Eq. (130) becomes

$$A_2(t) = \exp[- (1 - q^2)\langle \Delta\omega^2 \rangle \tau_s^2 f(t/\tau_s) - t/T_2^f], \tag{133}$$

where

$$(T_2^f)^{-1} = \langle \Delta\omega^2 \rangle q^2 \tau_f. \tag{134}$$

This expression for $1 - q^2 \approx 10^{-4} - 10^{-5}$ well explains an appearance of Gauss-like TMD in rubber and its transition to an exponential shape.

At higher temperatures ($T > T_g + 150$ K), when two components (Gauss-like and exponential) appear in rubber, and when TMD has a super-Lorentzian shape in thermoplastics, the relaxation function is to be derived from Eq. (131) that corresponds to a heterogeneous spin system. If anisotropy of fast motion is created by a network of physical or chemical crosslinks, a distribution of residual ineractions takes place in real polymers due to the heterogeneity of this network.

Let us assume that $(1 - q^2)\langle\Delta\omega^2\rangle$ takes on two values only, which are equal and unequal to zero. It means that there are chains inserted in the network and others that are not. For the first is $q^2 \neq 1$ and for the second $q^2 = 1$. The part of chain links with $q^2 \neq 1$ is designated as p. For the TMD shape follows a sum of two terms:

$$A_2(t) = p\exp\left[-\langle\Delta\omega^2\rangle(1-q^2)\tau_s^2 f\left(\frac{t}{\tau_s}\right) - \frac{t}{T_2^f}\right] + (1-p)\exp\left(-\frac{t}{T_2^f}\right). \quad (135)$$

One can see that Eq. (135) completely describes the evolution of TMD in elastomers [319]. The TMD can be characterized by three parameters p, $(1-q^2)\langle\Delta\omega^2\rangle$ and T_2^f, two of which are connected with the parameters of the chain inserted in the network of transversal connections (p and $(1-q^2)\langle\Delta\omega^2\rangle$) while the third ($T_2^f$) is determined by the behaviour of the free chain.

For thermoplastics, the fact that Gauss-like TMD shape and solid-echo signals are absent at high temperatures shows that the residual dipolar interaction is small, and the TMD can be described by the expression

$$A_2(t) = \sum_i p_i \exp\left(-\frac{t}{T_{2i}^f}\right) \quad (136)$$

that is obtained from Eq. (131) under the assumption of $(1-q^2)\langle\Delta\omega^2\rangle\tau_s \ll 1/T_2^f$. The comparison of Eqs. (135) and (136) shows that the nature of the exponential tail in rubber and the whole signal in thermoplastics is the same.

We have compared temperature dependences of spin-spin relaxation times measured in narrow fractions of the same molecular weight for atactic PS with PBD in order to test this conclusion. The quantities T_2^e characterizing the signal decay to e^{-1} of the initial amplitude and the decay time of the "tail" in PBD (T_2^f) were compared.

Figure 19 shows the results for the two pairs of samples with $M_w = 2 \times 10^4$ and 2×10^5. $T_2^{e, f}/T_2$ is represented as a function of the temperature difference $(T - T_g)$. (T_2 is the spin–spin relaxation time measured at a temperature a little lower than T_g.) One can see that in the temperature region $T \lesssim T_g + 70$ K the reduced values $T_2^{e, f}/T_2^{rl}$ coincide for all polymers. At higher temperatures T_2^e for PS coincides with T_2^f for PBD for corresponding molecular weights.

An assumption to the discrete distribution of residual dipole–dipolar interactions that are due to the chains inserted and not inserted into the network is quite insufficient to explain the line shape of a crosslinked polymer under stretching or swelling. In these cases we have to resort to the continuous distribution of residual

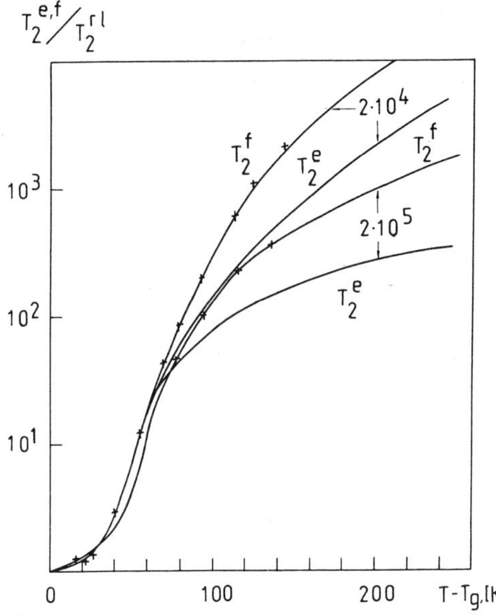

Fig. 19. Temperature dependence of the spin–spin–relaxation time in relative coordinates in fractionated PBD (solid lines) and in atactic PS (crosses) [60]. Parameter on the curves: molecular masses.

interactions which would be determined by the heterogeneity of the chemical network. In [142] a theoretical calculation of the line shape for crosslinked swollen polymers is performed using the model of flexible chains which have fixed ends. Expressions result that connect the proton resonance line shape with the molecular parameters of the polymer network for isolated spin pairs which are firmly fixed to the basic chain. The analysis showed that the super-Lorentzian line shape observed in the experiment [129, 138] can be obtained only on the assumption of a sufficiently broad distribution of the distances between cross-links.

3.1.5 Connection Between the Signal Parameters and the Polymer Network Structure

The formalism given above shows that the residual dipole–dipolar interaction is due to the forces which act on the polymer chains and limit their micro-Brownian (segmental) motion. It is obvious that these forces may be due to the fluctuating entanglements as well as to transversal chemical bonds. The latter in contrast to the first are independent of the temperature and do not break after swelling.

The parameters of TMD and the polymer network characteristics were connected quantitatively in [143]. Taking the model mentioned as the basis for this work one gets an expression for the residual dipole–dipolar interaction:

$$\langle\Delta\omega_s^2\rangle = \langle\Delta\omega^2\rangle\left(\frac{3}{5Z}\right)^2 = (1-q^2)\langle\Delta\omega^2\rangle \tag{137}$$

where $\langle\Delta\omega^2\rangle$ is the second moment measured at a temperature a little lower than

T_g, $\langle \Delta \omega_s^2 \rangle$ is the remaining part of the second moment after averaging by the fast motion, Z is the number of segments disposed between the crosslinks.

Z results in dependence on the molecular mass of the chain section between crosslinks (M_c) as follows:

$$Z = \frac{M_c}{M_0 S},\tag{138}$$

where S is the number of monomeric units in the hard segment (thermodynamic flexibility characteristics) and M_0 is the molecular mass of the monomer unit.

The connection between the relative intensity $(1-p)$ of a weakly damped TMD component in elastomers and M_c was stated in [144]. It is based on the intensity of this component determined by the number of chains left beyond the network. The crosslinking can take place only if the chain reaches the critical quantity M_c. Therefore the molecular mass of the sections left beyond the network can vary from zero to M_c. It is obvious to identify these sections with the end sections of the macromolecules. Assuming their random distribution we can consider the average molecular mass of the end section to be approximately equal to $(1/2)M_c$.

Hence,

$$(1-p)M_w = 2 \times \tfrac{1}{2}M_c,\tag{139}$$

where M_w is the molecular mass of the whole chain, $(1-p)M_w$ is the molecular mass of the end section, and factor 2 means that the macromolecule has two free ends.

Equations (137) and (139) yield information on the density of the polymer network by means of two independent methods based on TMD shape analysis.

Principally these two functions can be used to determine the concentration of the network entanglements. Here one must take into account that the parameters measured in the experiment depend not only on the static quantity $q^2 \langle \Delta \omega^2 \rangle$ but also on the correlation time τ_s of this network fluctuation.

A number of investigations of swollen polymers and solutions and of crosslinked and filled rubbers [145–147, 319, 320] made the quantitative estimation of polymer network parameters possible. The latter papers compare the results of the method mentioned with those of the routine procedures and show good agreement.

NMR studies of swollen crosslinked polymer gels can also show the rapid segmental motion and the spatial restrictions of motions by crosslinks [148]. The discussion of the super-Lorentzian lines allows the determination of the residual second moment as a direct quantitative measure of the motional restrictions induced by crosslinking in swollen networks.

3.2 Longitudinal Magnetization of Protons

The fact that polymer spin systems, even amorphous ones, are heterogeneous makes it necessary to carry out an analysis of the longitudinal magnetization decay (LMD) in detail in order to determine the spin–lattice relaxation times.

The present section pays great attention to the analysis of the experiments in the rotating frame (RF) and in the dipolar system. Experiments with PETP, PS, LPE will be discussed as typical examples.

3.2.1 Signal Decomposition in Amorphous Polymers and Spin–Dynamics Effects

Amorphous polymers in the solid state represent a proper model to clear up the role of spin–spin and spin–lattice relaxation processes in the formation of magnetization decays observed for various pulse sequences.

Numerous research projects showed that the longitudinal magnetization decay in solid amorphous polymers in the laboratory frame (LF) is always exponential and can be described by a single relaxation time.

Relaxation measurements of the dipolar reservoir provide a more complicated picture. Typical decays for solid amorphous PTEP are shown in Fig. 20a. This figure demonstrates that the decay consists of a fast and a slow decaying exponent. The first one is characterized by a time constant of 10^{-5} s and the second by 10^{-3} s.

An even more complicated picture is observed in the rotating frame (RF) in spin-locking experiments. One can see the typical experiment in Fig. 20b, which demonstrates clearly visible damped oscillations with a duration of 50–100 μs in the initial part of the decay. Frequency and amplitude of these oscillations depend on the magnetic induction B_1. Magnetization decreases monotonously after the oscillation process finished. From the experiments follows that the magnetization decay in the laboratory frame (LF) is described by a single exponent, in the dipolar system that is described by two and in RF by three components, which are one oscillating and two exponential.

Now the longitudinal magnetization is analyzed within the limits of the theory of "two reservoirs" for NMR in solids, the elements of which have been stated in Sect. 1.2.

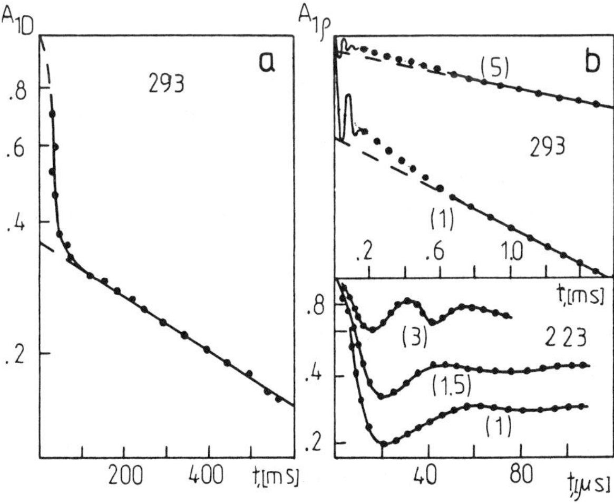

Fig. 20. Typical longitudinal magnetization decay in the dipolar system (a) and in the rotating frame (b) in amorphous PETP [60]. Parameter on the curves: temperatures in K, (b): B_1 in 0.1 mT, bottom: initial part of the decay.

Dipolar reservoir influence can be disregarded in LF experiments because the condition $B_0 \gg B_{loc}$ is always fulfilled and the decay is completely determined by the spin–lattice relaxation of the Zeeman reservoir characterized by the time T_1.

In the three-pulse Jeener experiment ([104], cf. Table 2) the effect of the first two pulses results in energy flow from the Zeeman to the dipolar reservoir. In a time t, which is of the order of T_2, the transversal component of magnetization vanishes in the system and a quasi-equilibrium occurs with a spin temperature which is determined by the dipolar system. The signal maximum that appears after the third pulse is inversely proportional to the spin temperature of the dipolar reservoir, and its decrease with the time characterizes the spin–lattice relaxation process. The fast decaying part of the signal is determined by the temperature establishing process while the slowly decaying part is connected with the spin lattice relaxation process. The fact that the relative intensity of the fast signal component is equal to 0.5, which is close to the theoretical coefficient (0.56) of magnetization transfer from the Zeeman to the dipolar reservoir proves that this component is the reflection of a spin–dynamics process.

Experimental magnetization decays in RF, shown in Fig. 20b are compared with the theoretical results (see Sect. 1.2.4), demonstrating that the theory provides a proper qualitative description of the experimental curves. In order to carry out a complete quantitative analysis in [36] field dependences of the intensities, of the relaxation times $T_{1\varrho i}$ of all components, and of the frequencies of the oscillating component with the theoretical data are compared in detail. Complete agreement with the theory was achieved.

Hence, the whole experiment can properly be described by the two-reservoir model of the solid spin system structure. One can assume that multicomponentness of longitudinal magnetization decays in amorphous solid polymers is due to the process of the establishment of the thermal equilibrium within the spin system and by no means to the complexity of the spin–lattice relaxation caused by the structural heterogeneity of amorphous polymers. The spin–lattice relaxation process in solid amorphous polymers is exponential and is characterized by a single relaxation time (T_1, $T_{1\varrho}$, T_{1Z}, or T_{1D}) in dependence on the experimental conditions.

It should be noted that this conclusion does not concern the decomposition of the spin-system into separate subsystems; neither the Zeeman (Z) nor the dipolar (D) reservoirs are described by a common spin temperature. In this case spin–lattice relaxation obeys the regularities of heterogeneous systems and is represented by a sum of several components. This case can only occur in amorphous polymers at temperatures higher than T_g when the transversal relaxation function is also multicomponent.

3.2.2 Analysis of Signals in Amorphous Polymers with Molecular Heterogeneities

Heterogeneities on the molecular level, mentioned before occur in polystyrene e.g. and give rise to a multicomponent decay of the relaxation function. Under proper conditions (high temperatures, small molecular mass fractions, and molecular masses smaller than the critical mass M_c) even a splitting of the lines can be observed in a normal FT-experiment without multiple-pulse technique or magic-angle sample rotation. Therefore, interesting experiments in polystyrene melt [42, 149] were performed for relatively narrow molecular mass fractions with weight-averaged molecular masses M_w from 2,100 up to 630,000 at temperatures more than 75 K higher than the glass transition temperature. Because of the presence of two lines in the spectrum, split by approx. 5 ppm and well separated for molecular masses smaller than M_c (cf. Fig. 21), T_1 inversion recovery measurements at 88 MHz were possible not only in a usual non-selective manner, but also selectively using the FT-technique (linewidth approx. 2 ppm). The high resolution of the spectra without additional experimental technique is firstly due to the averaged dipolar interaction caused by the fast reorientations in the polymer melt at temperatures sufficiently high compared to the glass transition. Secondly, the hindrances of the rearrangements by entanglements are negligible because the free volume is big for the small molecular masses. The line width, of course, increases at lower temperatures and for higher molecular masses. For molecular masses smaller than M_c the non-selective T_1 measurements show two components with an intensity ratio of $3:5$ according to the proton number in the polymer chain and the

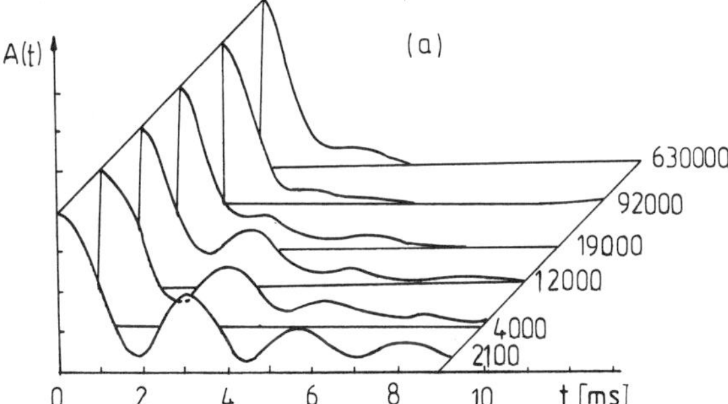

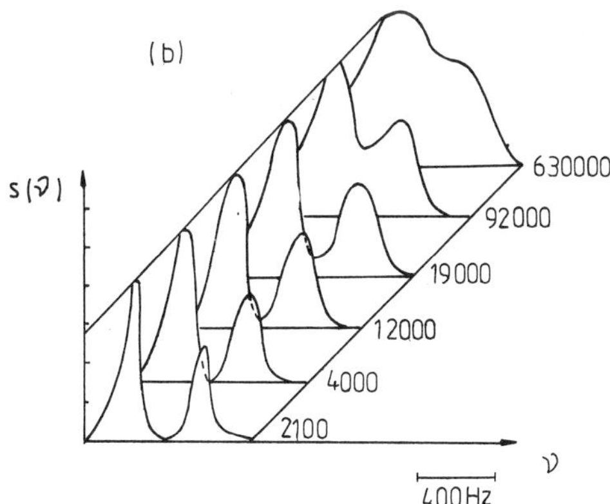

Fig. 21. Proton FID signals (a) and Fourier transformed spectra (b) of atactic PS at 493 K (measuring frequency 88 MHz, off-resonance 5 KHz). Parameters on the curves: molecular masses

phenyl rings. The higher T_1 values of the aromatic protons are due to the small second moment in comparison with the moment of the aliphatic protons.

The FT-inversion recovery experiments yield the same results with regard to the intensities as well as to the relaxation times. This good accordance verifies the correctness of the assignment of the components, which was done during the decomposition of the relaxation function. The increase of the intensity of the long component shows, for higher molecular weights, that the phenyl groups represent relaxation centres for the chain protons because of their higher mobility. The temperature dependence of both T_1 components shows broad minima, which can be explained by broad distributions of correlation times using the WLF-equation, which is characteristic for a large cooperativity of the motions. The T_1 minimum for the aromatic protons at lower temperatures once more shows the higher molecular mobility of the phenyl protons. Finally the shift of the minima to higher temperatures with increasing molecular masses reflects the decrease of the free volume in the same way as the line broadening.

3.2.3   Signal Analysis in Semicrystalline Polymers

According to the results of TMD analysis the spin system of partially crystalline polymers consists of three subsystems corresponding to the three structural polymer phases. Since these phases are disposed in immediate contact with each other by the passage of some macromolecules through all three phases, the interaction that leads to magnetization transfer between the subsystems can take place along the boundary layers of these phases.

The hypothetic spin system and possible relaxation processes of three-phase polymers are schematically shown in Fig. 22. The designation are the same as in Fig. 3. In addition, K_{ij} are the rates of the interphase spin diffusion, indices i, j designate the attribution of the spin system parameters of the phases (cr, int, and am). Each phase is characterized by the local field and by the relative number of spins P_i.

The longitudinal magnetization decays in different ways depending on the parameters K_{ij}, T_{1i}, T_{1Di}, T_{1Zi}, and P_{1i} and on the experimental conditions. Under certain conditions ($K_{ij} \gg T_{1i}^{-1}$, T_{1Di}^{-1}, T_{1Zi}^{-1}) the magnetization decays purely exponentially. Under other conditions (e.g. $K_{ij} \ll T_{1Di}^{-1}$, T_{1Zi}^{-1}, and $B_1 \approx B_{loc}$) the number of components can be up to six.

We shall study some typical situations that occur in longitudinal relaxation analysis under various experimental conditions using the two polymers PE and PETP.

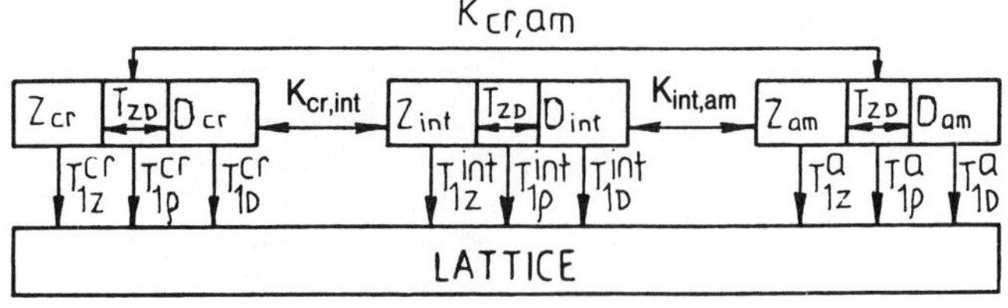

Fig. 22. Hypothetic spin system of a polymer including three phases

3.2.3.1 Linear Polyethylene

A) Laboratory Frame

Our experiments [150, 116] as well as those described in [151, 152] show that longitudinal magnetization decays in LF for linear PE are non-exponential in the minimum region (240–300 K) and exponential at all other temperatures. We carried out special experiments [116] with the schemes given in Section 2.2 in order to clear up the nature of this nonexponentiality. Nonexponentiality of LMD has turned out to be due to the fact that the TMD consists of three components and hence, due to the phase structure. The short (crystalline) FID component corresponds to the long LMD component while the long (amorphous) FID component corresponds to the short LMD component.

B) Dipolar System

Our paper [153] devoted to dipolar echo decay shape examination shows that the LMD in spin–lattice relaxation T_{1D} measurements by the Jeener method is always described by a sum of three exponents (excluding the very rapidly decaying initial part) with decay rates which differ from one another by the factors 5–10. The dipolar echo (DE) signal shape with various intervals between the second and the third pulses was examined to clear up the nature of these components. Figure 23 shows the typical result of such an experiment. One can see that the DE shape contains all three components for $t < T_{1D3}$, that the amorphous component vanishes for $T_{1D3} < t < T_{1D2}$, that the signal DE consists of a single crystalline component only for $T_{1D2} < t < T_{1D1}$. This means a complete qualitative correspondence of longitudinal magnetization components and PE structure phases. Such correspondence was proved in the temperature interval from 130 to 320 K. At higher temperatures $T > T_g^{am}$ LMD becomes two-component since the component corresponding to the amorphous phase is not observed in the experiment. It

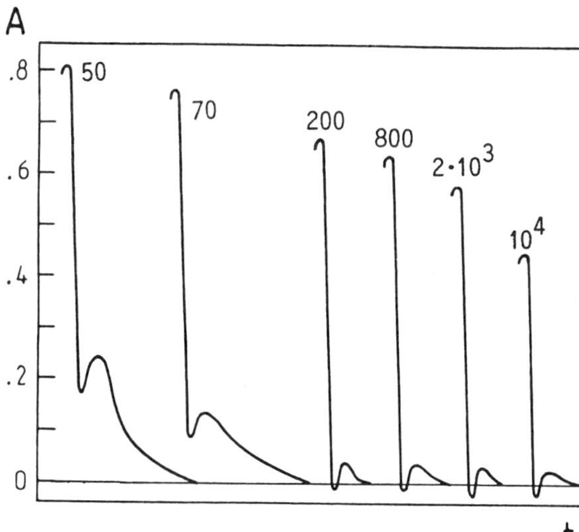

Fig. 23. Free induction decay after the third pulse in a T_{1D}-experiment (cf. Table 2) in LPE at 303 K [153]. Parameters on the curves: t_1 in μs.

happens due to the very small local field of the amorphous phase and because an effective magnetization transfer to the dipolar system of the amorphous phase does not take place under the affection of the first two pulses. This component relaxes with the time T_2^{am}.

C) Rotating Frame

We carried out all experiments in RF for PE [115] in a field $B_1 = 4mT$. Figure 24 shows the typical longitudinal magnetization decay in RF that is observed in PE. It is clear that the signal can be formed by a sum of several components

$$A(t) = P_{osc} f_{osc}(t) + \sum_i P_{1\varrho i} \exp\left(-\frac{t}{T_{1\varrho i}}\right), \tag{140}$$

where $f_{osc}(t)$ is a function describing the damped oscillations, P_{osc} is the corresponding relative intensity, $P_{1\varrho i}$ are the intensities of the signal components relaxing

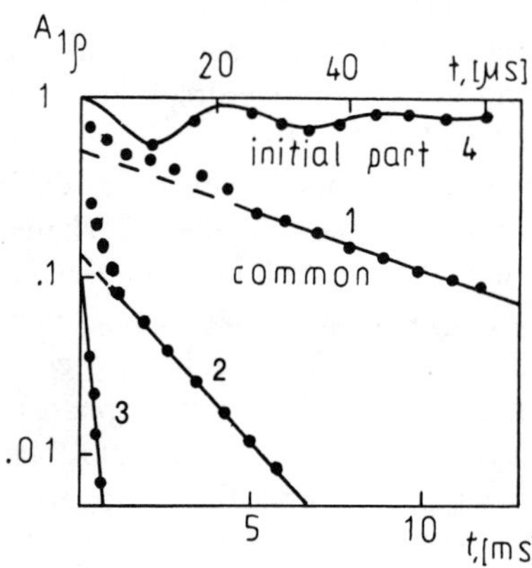

Fig. 24. Typical longitudinal magnetization decay in the rotating frame in LPE [60]

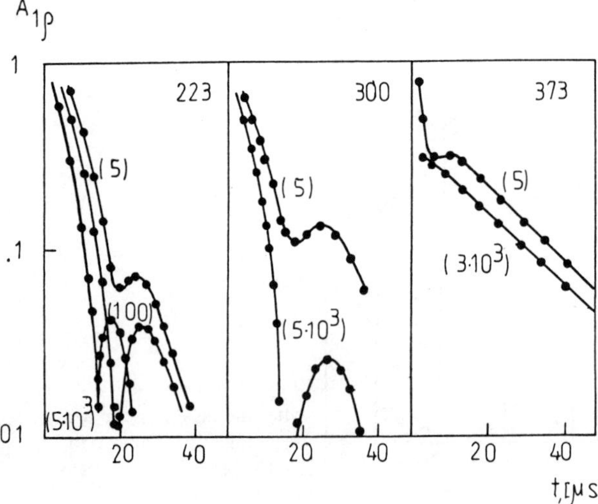

Fig. 25. Free induction decay after the long *rf*-pulse in a spin-locking experiment (cf. Table 2) in LPE at different temperatures. Parameters on the curves: temperatures in K, in parentheses: t_1 in μs

with the times $T_{1\varrho i}$. The number of exponential components at temperatures lower than 220 K was four and at temperatures higher than 220 K three. Component numeration begins from the component which is decaying most slowly. We used exactly the same methods of two-dimensional relaxation analysis to attribute components to phases. Decays of longitudinal magnetization that are observed for various durations of the locking pulse for three characteristic temperatures are demonstrated in Fig. 25.

Detailed analysis of this experiment carried out in [115] shows:

(i) There are four temperature regions with different correspondence of longitudinal and transversal magnetization components.
(ii) One part of the components is determined by spin–lattice interaction, another by spin–spin interaction.
(iii) The experiment can be described properly assuming two subsystems (cr and am).
(iv) Spin–lattice relaxation processes determine the signals of the longest relaxation times in each spin system.

Table 5 shows the attribution of signal components in RF and calculation formulas for the intensities of corresponding spin systems in various temperature intervals.

Table 5. Attribution of the measured intensities to the populations of structural phases in LPE

$T[\mathrm{K}]$	< 200	$200-300$	$330-360$	> 360
$P'_{1\varrho\,\mathrm{cr}}$	$P_{1\varrho 1} + P_{1\varrho 2} + \frac{1}{2}P_{\mathrm{osc}}$	$P_{1\varrho 1} + P_{1\varrho 2} + P_{\mathrm{osc}}$	$P_{1\varrho 1} + P_{1\varrho 3} + P_{\mathrm{osc}}$	$P_{1\varrho 2} + P_{1\varrho 3} + P_{\mathrm{osc}}$
$P'_{1\varrho\,\mathrm{am}}$	$P_{1\varrho 3} + P_{1\varrho 4} + \frac{1}{2}P_{\mathrm{osc}}$	$P_{1\varrho 3}$	$P_{1\varrho 2}$	$P_{1\varrho 1}$

Temperature dependences of the relative intensities (P_2, P'_1, $P'_{1\varrho}$, and P_{1D}, $'$ means apparent quantities) for the same sample of linear polyethylene are shown in Fig. 26. The figure demonstrates that the intensities P_{1D} for all three components agree quite well with the intensities P_2. The intensities $P'_{1\,\mathrm{am}}$ and $P'_{1\varrho\,\mathrm{am}}$ are close or equal to the intensities $P_{2\,\mathrm{am}}$, while $P'_{1\varrho\,\mathrm{cr}}$ and $P'_{1\,\mathrm{cr}}$ are close to the sum of the two intensities $P_{2\,\mathrm{cr}} + P_{2\,\mathrm{int}}$ of the transversal magnetization.

The intensities in LF are a little different from those obtained in other experiments and depend on the temperature. The complete correspondence of the intensities P_{1D} and P_2 shows that the rates of magnetization transfer between the phases are smaller than the rates of spin–lattice relaxation T_{1D}^{-1} and the conditions of slow exchange are fulfilled. This also shows that the relaxation of all three systems to the lattice is independent from one another and that the relaxation times are true, i.e. not distorted by interphase exchange.

The correspondence of the quantitites $P'_{1\varrho\,\mathrm{am}}$, $P_{2\,\mathrm{am}}$ and $P'_{1\varrho\,\mathrm{cr}} = P_{2\,\mathrm{cr}} + P_{2\,\mathrm{int}}$ implies that there is a rapid magnetization transfer between the spin systems of crystalline and intermediate phases, which joins them together in a common spin system, relaxing with weight averaged rate of $T_{1\varrho\,\mathrm{cr+int}}^{-1} = P_{\mathrm{cr}} T_{1\varrho\,\mathrm{cr}}^{-1} + P_{\mathrm{int}} T_{1\varrho\,\mathrm{int}}^{-1}$. At the

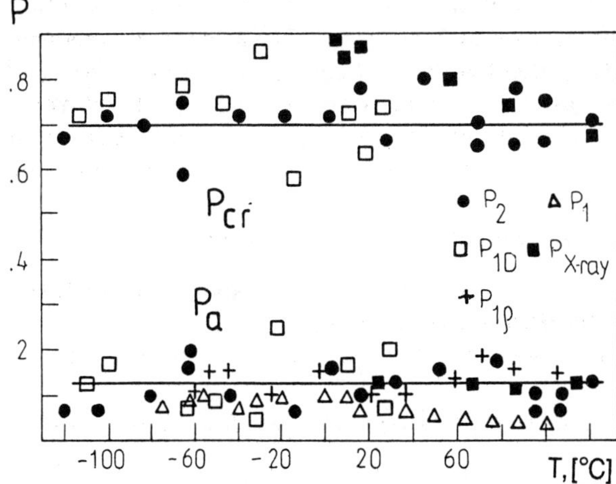

Fig. 26. Temperature dependence of the intensities of different pulse-NMR and X-ray experiments in LPE ($M_W \approx 3 \times 10^5$) [60]

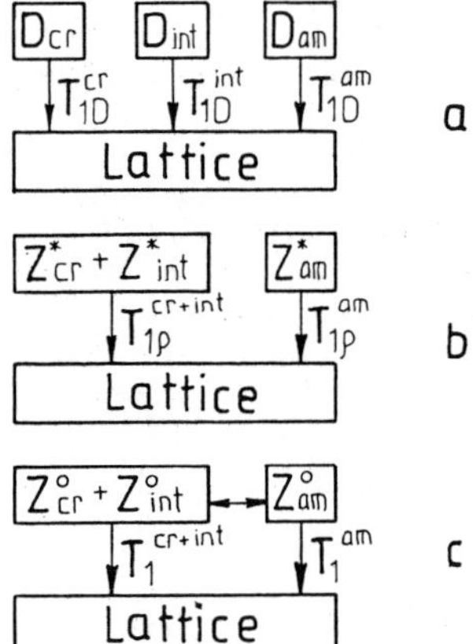

Fig. 27. Schemes of effective spin systems for **(a)** T_{1D}-, **(b)** $T_{1\rho}$-, **(c)** T_1-experiments in LPE [60]

same time this joined system and the system of the amorphous phase are connected by a slow exchange condition that leads to an independent relaxation of each spin system.

Though the construction of the spin system is the same as in RF in general the correspondence of the components in LF signifies, that the magnetization transfer with rates comparable with those of the spin–lattice relaxation takes place between the joined ($cr + im$) and amorphous systems and that the relaxation times measured are apparent only, i.e. they depend on the transfer rates. Figure 27 shows typical spin system structures in PE obtained in different experiments.

3.2.3.2 *Poly(ethylene terephthalate)*

The papers [38, 37, 60] analyse in detail multicomponent magnetization decays in a series of semicrystalline PETP of different crystallinity that had been achieved by thermal pretreatment of the samples. Experiments for the proper attribution of LMD and TMD components resulted in the construction of effective spin systems. The analysis was generally made according to the same principle as for polyethylene. In this investigation we succeeded, firstly, in stating that the observed multicomponentness of signals is completely due to the polymer phase structure and not at all to the structural heterogeneity of its molecular chain and, secondly, the possibility of explaining the whole data within the limits of the three-phase model of the polymer structure was shown.

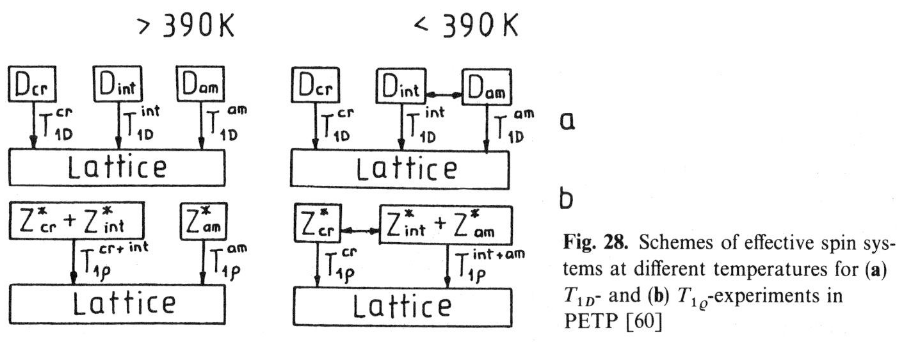

Fig. 28. Schemes of effective spin systems at different temperatures for (**a**) T_{1D}- and (**b**) $T_{1\varrho}$-experiments in PETP [60]

The construction of an effective spin system for PETP ($\varepsilon = 0.55$), shown in Fig. 28, demonstrates that they vary not only with the type of experiment, but also with the temperature. From Fig. 28 one can see that the dipolar system consists of three unconnected subsystems at temperatures higher than 390 K, while at temperatures lower than 390 K it consists of a single independent (crystalline) and two partially connected subsystems. In the first case the relaxation times determined in the experiment are true for each phase. At lower temperatures only the times corresponding to the crystalline phase are true while T_{1D2} and T_{1D3} depend on the magnetization transfer rates. At temperatures lower than 220 K magnetization transfer rates between amorphous and intermediate phases become so great that the spin systems are united and relax to a lattice with a weight averaged time.

At temperatures lower than 390 K the RF spin system decomposes into two systems corresponding to the crystalline and the united intermediate and amorphous systems; they are independent within a temperature interval of 230 to 390 K. At lower temperatures magnetization transfer becomes appreciable, and measured relaxation times are only apparent.

This picture sharply changes at temperatures higher than 390 K: the spin systems of crystalline and intermediate phases unite and form a common spin system, while the amorphous phase becomes independent.

Figure 29 shows a typical temperature dependence of the measured intensities for PETP.

Cheung et al. [120] use the combination of multiple pulse sequences for line narrowing and sample rotation under the magic angle and obtain high resolution

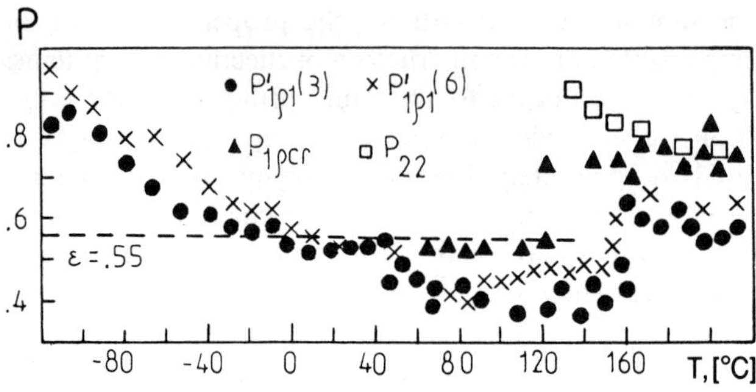

Fig. 29. Temperature dependences of the relative intensities of different pulse experiments in PTEP with a degree of crystallinity of 0.55 [38]. In parentheses: B_1 in 0.1 mT. $P_{1\varrho cr}$ obtained from experimental values $P_{1\varrho1}$ in an analogous manner as demonstrated in Tab. 5 for LPE.

spectra in solid crystalline PETP. The spin–lattice relaxation times were separately measured in RF for methylene and aromatic protons. In both cases decays were described by two exponents with relaxation times similar for both proton species. Hence, the two spin–lattice relaxation components correspond to two structural phases but not to the two proton species (methylene and aromatic) in the sample. The interphase spin diffusion process proceeds between two spatially separated phases, and the fast spin diffusion within each phase unites the spin systems of two different types of protons even under high resolution conditions.

These conclusions concerning the spin system structure in PETP and the dynamic processes taking place in the latter were obtained by means of non-selective pulse experiments and completely confirmed by results obtained in multiple-pulse experiments.

3.2.3.3 Quantitative Calculation of Spin-Diffusion Effects

We employ the simplest model of a two-phase system in order to calculate the influence of magnetization transfer on the measured parameters of individual spin systems quantitatively. Kinetic equations describing the magnetization transfer from one spin system to another, provided that the temperature is established instantaneously within each spin system ($T_2 \ll T_{1,1\varrho,1D}$), are similar to the well-known McConnel equations [31, 154, 155] taking the chemical exchange into account. The investigations [37, 155] show that this approach is realized in nuclear relaxation analysis for partially crystalline polymers.

The solution of these equations obtained from [156] shows the longitudinal magnetization decay as a sum of two exponential terms

$$A(t) = P'_{11} \exp\left(-\frac{t}{T'_{11}}\right) + P'_{12} \exp\left(-\frac{t}{T'_{12}}\right), \tag{141}$$

where $T'_{11,12}$ and $P'_{11,12}$ are functions of the exchange rates $K_{1,2}$, and of the true relaxation times and intensities.

The analysis of these functions shows that if the transfer rates (K) are comparable with the relaxation rates ($1/T_1$), only the apparent parameters of the spin system are measured in the experiment. In the case $K \ll 1/T_1$ the true parameters, i.e. uninfluenced by transfer, are measured in the experiment.

As an example we shall consider an experiment in RF in PETP [37, 38]. Figure 29 shows the temperature dependences of the measured intensities $P'_{1\varrho l}$. So a plateau is observed at medium temperatures while for increasing as well as for decreasing temperatures $P'_{1\varrho}$ grows. The quantity measured on the plateau agrees with sufficient exactness with the crystallinity of each sample. Hence, one can conclude that the true intensity is measured on the plateau and corresponds to the crystallinity, i.e. the condition of slow interphase exchange is fulfilled and the true relaxation times of the crystalline and of the united amorphous and intermediate phases are measured in the experiment.

We can now calculate the true system parameters from the experimental values $T'_{1\varrho cr}$, $T'_{1\varrho am}$ and $P'_{1\varrho cr}$ on the assumption that $P_{1\varrho cr} = \varepsilon$ is constant for the whole temperature region. We shall do this with the help of the expressions (cf. [156]) derived for true parameters.

The examples of temperature dependences of true and apparent (experimental) relaxation times $T_{1\varrho}$ for PETP and T_1 for PE are represented in Fig. 30. The procedure to obtain the true relaxation times for PE is similar and is described in detail in [116].

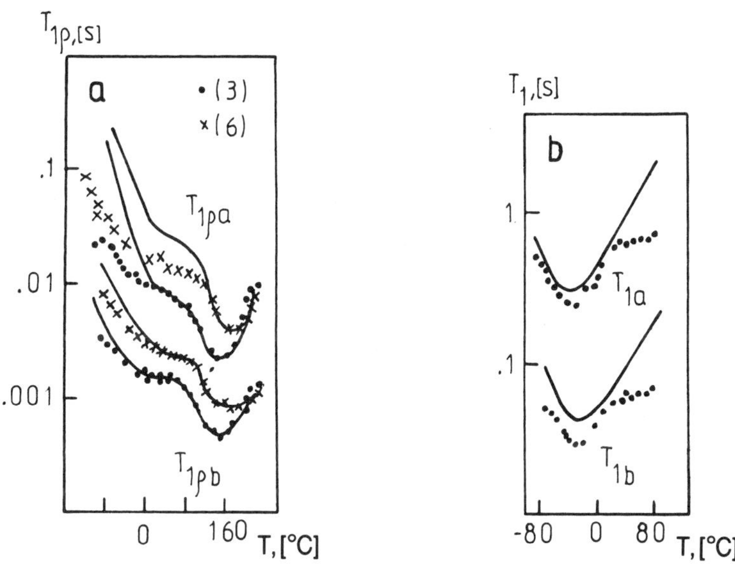

Fig. 30. Temperature dependences of measured (points) and calculated according to [156] (solid lines) relaxation times (**a**) $T_{1\varrho}$ in PETP [37, 38] and (**b**) in LPE [116]. In parentheses: B_1 in 0.1 mT

Figure 31 represents the temperature dependences of exchange rates (K) obtained from the experimental data analysis for several PETP samples. The fact that the exchange rates increase with decreasing temperature in the region of low temperatures in all samples leads to the conclusion that spin diffusion is the transfer mechanism. At the same time the rise of K with temperature observed at high temperatures in the PETP sample of the least crystallinity cannot be described by

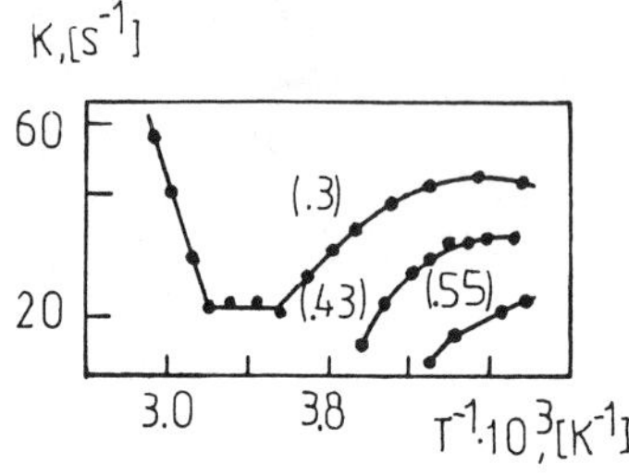

Fig. 31. Temperature dependence of the exchange rate calculated from apparent relaxation times for PETP [38]. In parentheses: degree of crystallinity

the same mechanism. This leads to the assumption that a material exchange that may be caused by segment motion in the contact zone here takes place between phases. The rise of crystallinity accompanied by decreasing exchange rates is most likely caused by the increase of the specific surface of the crystallites that are in contact with the amorphous phases. This fact correlates with the rise of the size of the crystallites that is observed with the increase of the annealing temperature of PETP. The insufficient accuracy of the analysis of the experimental data makes it impossible to reveal exactly the temperature dependence of K and, hence, the exchange mechanism. It only becomes clear that the exchange rates in PE are much smaller than those in PETP (by two orders of magnitude); this fact reflects phase structure peculiarities.

A more direct method permitting the determination of the rates of magnetization transfer between different polymer structures, the so-called Goldman–Shen experiment (cf. Table 2) has advanced recently. This method was used to examine the domain structure in block and segment copolymers [118] and the structure of the amorphous regions in crystalline polymers (PE, PP, PETP) [119]. These investigations show ways to obtain information on the size of the structure regions from the constants of spin diffusion.

3.2.4 Comparative Analysis of the Experiments on Pulse and Continuous Spin-Locking

In this section we will provide a comparative analysis of the results of the experiments on pulse and continuous spin-locking in order to state the limits of the applicability of the MW4 pulse sequence (cf. Table 2) to the spin–lattice relaxation time measurements in different experimental situations.

The temperature and field dependences of $T_{1\varrho}$ and T_{2e} in amorphous polymers were investigated at two typical temperature ranges. These were an α-transition region (liquid-like segmental motion) and a β-transition region (local anisotropic motion). Here we will discuss the results of our experiments on PETP [157, 60], PS [60], and PVC [40]. To compare the two experiments we reduced the pulse distance 2τ to the averaged effective field $B_{1\,\text{eff}}$ that affected the sample spin system by a series of 90° pulses. This quantity corresponds to B_1 that is the field of the spin-locking experiment. The minima characteristic for the spin–lattice relaxation observed at the temperature dependences T_{2e} and $T_{1\varrho}$ are due to the same reasons.

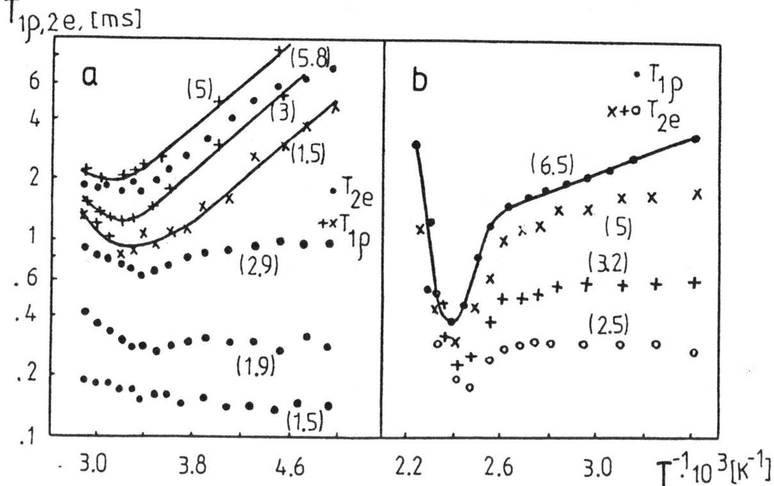

Fig. 32. Temperature dependence of $T_{1\varrho}$ and T_{2e} for (**a**) PETP and (**b**) PVC. In parentheses: B_1 or $B_{1\,eff}$ in 0.1 mT. Solid lines: Fitting of $T_{1\varrho}$ data

Table 6. Relaxation times $T_{1\varrho}$ and T_{2e} at the minima and their ratios for some amorphous polymers [60]

B_1 [mT]	Local motion PETP $T_{1\varrho}$ [ms]	T_{2e} [ms]	$T_{2e}/T_{1\varrho}$	Segmental motion PS $T_{1\varrho}$ [ms]	T_{2e} [ms]	$T_{2e}/T_{1\varrho}$	PVC $T_{1\varrho}$ [ms]	T_{2e} [ms]	$T_{2e}/T_{1\varrho}$
0.25	1.8	0.5	0.28	—	1.8	—	—	2.0	—
0.35	2.2	1.2	0.55	—	2.2	—	—	2.9	—
0.50	3.2	2.5	0.78	—	3.0	—	3.4	4.1	1.2
0.60	4.0	3.5	0.88	3.5	4.0	1.14	—	5.0	—

At low temperatures (lower than the minimum temperature) the behaviour of T_{2e} and $T_{1\varrho}$ differs sharply (Fig. 32). So $T_{2\mathrm{eff}}$ reaches a plateau while $T_{1\varrho}$ increases continuously with the temperature decrease. The position of the T_{2e} plateau depends strongly on the $B_{1\,eff}$ field. On the plateau is $T_{2e} \sim B_1^{2.5-3}$ while $T_{1\varrho} \sim B_1^{0.8}$. In the region of the minima the dependence of $T_{1\varrho}$ and T_{2e} on the field B_1 is approximately linear. Table 6 presents the relaxation times measured at the minima and their ratios obtained in various B_1 fields.

The results of the experiment are interpreted using the theory that describes the effect of the multiple-pulse sequences on the spin systems of solids in the absence of motion [43] and at slow motion [48].

According to [48] the expression for $1/T_{2e}$ consists of two parts:

$$\frac{1}{T_{2e}} = \frac{1}{T'_{2e}} + \frac{1}{T''_{2e}},$$

(142)

where $1/T'_{2e}$ is the decay rate in a solid lattice, and

$$\frac{1}{T''_{2e}} \approx \frac{\pi^2}{8} \frac{1}{T_{1\varrho}} \tag{143}$$

is due to the motion.

Table 6 shows that the ratio of the relaxation times at the minimum for segmental motion (PS and PVC) is equal to $\pi^2/8$ even at sufficiently rapid motions ($\tau_c \approx T_2$) and hence, $1/T'_{2e} = 0$ (cf. Sect. 1.2). For the local motions in PETP $1/T_{2e}$ contains the $1/T'_{2e}$ term and the measured relaxation rate is not equal to the rate of spin–lattice relaxation $T_{1\varrho}^{-1}$. The term $1/T'_{2e}$ decreases sharply during B_1 growth and approaches zero at 0.8–1 mT. Assuming the correctness of Eq. (142) it is possible to estimate the field dependence of T'_{2e} on the plateau, which in our estimation is equal to $B_1^{4.0}$ for PS and to $B_1^{3.2}$ for PETP, which is quite close to the theory [43].

The effect of the phase heterogeneity on the results of the pulse spin-locking experiment was studied using the example of solid polyethylene.

The experiment was carried out with pulse distances τ of 6, 8 and 10 μs. The analysis showed that the magnetization decays can be described quite satisfactorily by a sum of two exponents at $\tau = 6\ \mu$s, while at $\tau = 8\ \mu$s and $\tau = 10\ \mu$s the decays are sometimes a sum of two and sometimes of three exponents.

The correlation of these decay components with the FID components shows that the short relaxation time is generally determined by the relaxation of the crystalline regions, while the long relaxation time is determined by the relaxation in the amorphous phase. Thus, the attribution of the components very often is quite opposite to that of the continuous spin-locking experiment.

Figure 33 represents the temperature dependences of T_{2e}^{am} and T_{2e}^{cr} together with those of $T_{1\varrho}^{am}$ and $T_{1\varrho}^{cr}$ [60]. An evident discrepancy of the results of pulse and of continuous spin-locking is due to the fact that the contribution of the spin dynamic

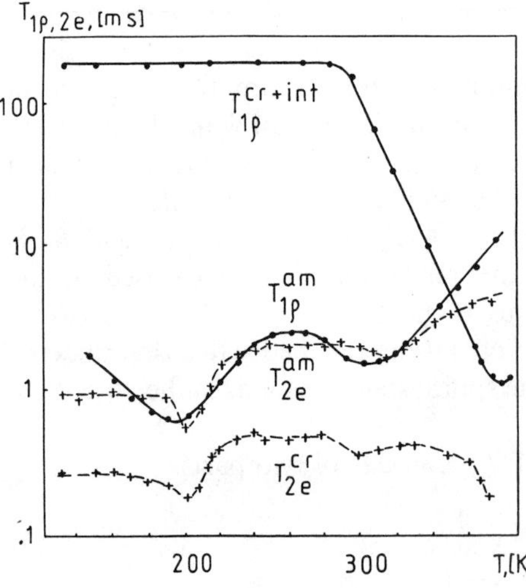

Fig. 33. Temperature dependence of $T_{1\varrho}$ and T_{2e} for LPE at $B_1 = 0.4$ mT. $B_{1\,eff} = 0.5$ mT

process $T_{2e}'^{-1}$ to T_{2e}^{-1} dominates in crystalline regions while in amorphous ones the contribution of this process is large only at very low temperatures. At temperatures higher than 200 K this contribution decreases and a quantity close to $T_{1\varrho}$ is measured in the experiment.

Hence, one can conclude that the use of the MW4 sequence for the measurements of spin–lattice relaxation determined by molecular motion is correct for sequential motion investigations, while for the investigations of local motions one must bear in mind the additional term.

In this case the measurements must be carried out only at high pulse repetition rates. Unfortunately the experimental possibilities of obtaining the necessary frequencies are limited.

3.3 Line Shape Analysis for ^{13}C- and ^{2}H-Spectra

In this section we shall consider the shape of relatively wide lines caused by anisotropy of chemical shift, dipolar or quadrupolar interaction, morphology or orientation of the sample (cf. e.g. [158–161]). Therefore it is very important to separate the individual lines, but nevertheless to preserve the interaction under consideration. For these applications we have already described 2D-spectroscopy, and in this chapter we shall discuss several methods of regaining the anisotropy of the chemical shift in MAS-experiments. Conclusions are drawn as to crystallinity, crystalline and amorphous domains, angles and distributions of orientations. The polymer microstructure, however, will primarily be discussed in Sect. 3.5, where the line positions in highly resolved spectra are considered.

3.1.1 Shielding Tensor Distribution in Isotropic Polymers

In Sect. 1.1.2 a mathematical expression was given for the line shape caused by chemical shift anisotropy with shielding tensor element distributions. There are many reasons for these distributions, which lead to the smoothing of the curves theoretically derived for pure polycrystalline materials [13–15, 162]. Consequently, the inverse way for the determination of these distribution functions from the experimental line shapes must in principle exist. The results of these distributions are well demonstrated in Fig. 34 showing the line shape for high-density polyethylene with a degree of crystallinity of approx. 60% and 85%, respectively, determined by X-ray scattering. For the sample with the higher crystallinity the spectrum becomes similar to the line shape actually occurring in a powder spectrum.

A) Reasons for the Distributions
Some of the influences which lead to a broadening of the spectrum can be excluded by an appropriate choice of the experimental conditions. This is the case for molecular motion, residual interactions as a consequence of incomplete dipolar decoupling, inhomogeneity or instability of the magnetic field, influence of the

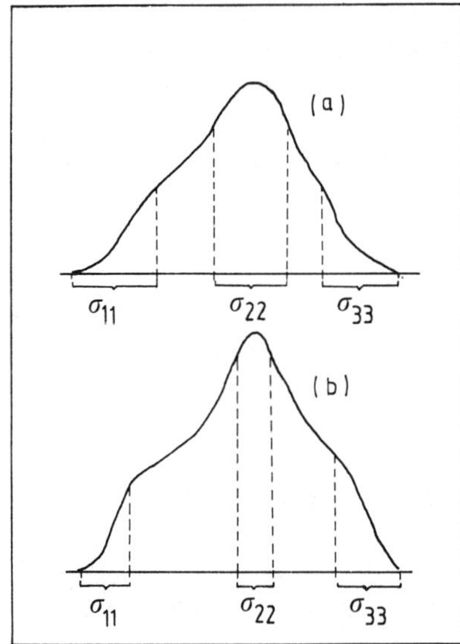

Fig. 34. ^{13}C cross-polarization spectra of polyethylene with degrees of crystallinity 60% (**a**) and 85% (**b**)

susceptibility caused by the geometry of the sample, and the ^{13}C–^{13}C interaction (which can reach approximately 4 ppm for polymers). The removal of these perturbations takes place to a certain extent by the application of lower temperatures, higher *rf*-fields, better stabilization and shimming of the static magnetic field, different sample preparation and an additional WHH4-pulse sequence on the ^{13}C frequency channel during the signal detection, respectively. The sum of all these broadenings is reduced to a value of about 1 ppm and does not influence the line shape. The main reason for the σ-distribution is the influence of the intermolecular structure on the chemical shift of the nuclei observed. The Van der Waals interaction is due to the influence of the neighbouring molecules on the deformation of the electron cloud surrounding the observed nucleus and leads, for the structural disorder in amorphous regions, to a spread of chemical shifts, which does not exceed 1 ppm. In solids a different shielding effect dominates. This is the shielding of the magnetic field on the site of the observed carbon by the polarized electron clouds of neighbouring nuclei if the distance is not too large. This problem can be treated by the point-dipole approximation for electron clouds suggested by McConnel [163]. However, as spheric symmetry is assumed in this method, the following equation for the contribution $\Delta\hat{\sigma}_{ij}$ of the electron cloud j to the chemical shift of the nucleus i is only valid, if the distances are not less than about 2 to 3 times the dimension of the cloud:

$$\Delta\hat{\sigma}_{ij} = \hat{\sigma}_{0j}\left(\frac{a}{r_{ij}}\right)^3 \frac{3}{2}(3\cos^2\theta_0 - 1), \tag{144}$$

where $\hat{\sigma}_{0j}$ is the chemical shift of the nucleus j in the centre of the electron cloud j, θ_0 is the angle between the vector $\vec{r}_{ij}$ and the static magnetic field, r_{ij} the magnitude of $\vec{r}_{ij}$, and a is Bohr's radius. Therefore, a more detailed discussion of the character of the distributions of the chemical shift tensor elements has to consider the latter intermolecular effects. That will be described in Sect. C (iii).

The differences in chain conformations influence the ^{13}C line widths not only in the anisotropic case, but also when magic angle spinning is applied. The remaining line width for glassy polymers of about 100 Hz arises not from instrumental limitations but from characteristics of the sample [14, 164, 165]. Using the rotational isomeric state model the chain conformations characterized by the position of the γ-carbons (those carbons that are three bonds away from the carbon being observed) are specified by *trans, gauche* +, and *gauche* −, respectively. This γ-effect [166, 167] causes an upfield shift for the observed carbon of approximately 3.5 ppm if one γ-carbon changes its conformation from *trans* to *gauche*. Furthermore, in [147, 165] various reasons for a line broadening in glassy, semicrystalline and crystalline polymers are discussed, whereby a substantial contribution arises from distributions of anisotropic sources of magnetic susceptibility, bond angles, or frozen molecular conformations.

B) Evaluation of the Distribution Function

In Eq. (29) the upper limit of the integration σ_{11}^u can be replaced by σ, since in the integrand

$$I(\sigma, \sigma_{11}, \sigma_{22}, \sigma_{33}) = 0, \quad \text{if} \quad \sigma < \sigma_{11}. \tag{145}$$

If the spread of the σ_{ii}-distribution is so small (cf. Fig. 35) that it provides no contributions in the regions of the other two σ_{jj}-distributions $(i \neq j)$, then e.g. the left flank of the spectrum is described only by the σ_{11}-distribution and the integrations over σ_{22} and σ_{33} yield the effective values σ_{2e} and σ_{3e} in $I(\sigma, \sigma_{11}, \sigma_{2e}, \sigma_{3e})$ in agreement with the mean value theorem of integral calculus. Therefore, for the determination of the distribution function $k_1(\sigma_{11})$ the following integral equation must be solved:

$$G(\sigma) = \int_{\sigma_{11}^l}^{\sigma} k_1(\sigma_{11})I(\sigma, \sigma_{11}, \sigma_{2e}, \sigma_{3e})d\sigma_{11}. \tag{146}$$

For this special type of integral the inversion is possible numerically. Because the parameter σ occurs in the integrand as well as in the upper limit, one can calculate

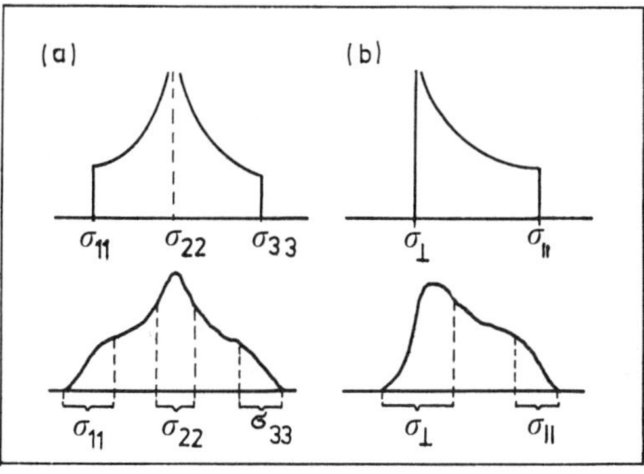

Fig. 35. Smoothing of the line shape in polymers because of chemical shift distribution for general (**a**) and axial (**b**) symmetry. Upper curves: Theoretical powder patterns, lower curves: smoothed spectra

$k_1(\sigma_{11})$ numerically step by step starting with $\sigma = \sigma_{11}^l$, where the necessary values for $k_1(\sigma_{11})$ in the respective step are already known from the former ones.

Unfortunately, the distribution functions still calculated in this way often contain contributions of other interactions, if the conditions for suppressing other mechanisms are not quite fulfilled such as in Hempels work [13, 15], where the determination of the distribution functions was done without $^{13}C–^{13}C$ decoupling. In this case it is necessary to perform a reconvolution of the distribution function with an appropriate choice of the convolution function (often of a Lorentzian shape), an estimation of the line shape from other considerations tolerating the disadvantage of decreasing the signal-noise ratio.

C) Applications

In [162, 13, 15] shielding tensor element distribution functions were measured and calculated for polycarbonate, polystyrene, polyethylene with different degrees of crystallinity, and, for comparison, hexamethylbenzene. We shall interpret these distribution functions (i) by comparing their width at half-height with the degree of crystallinity, (ii) by their decomposition to distinguished regions of different order in the polymer, and (iii) by studying the abundance of various conformations.

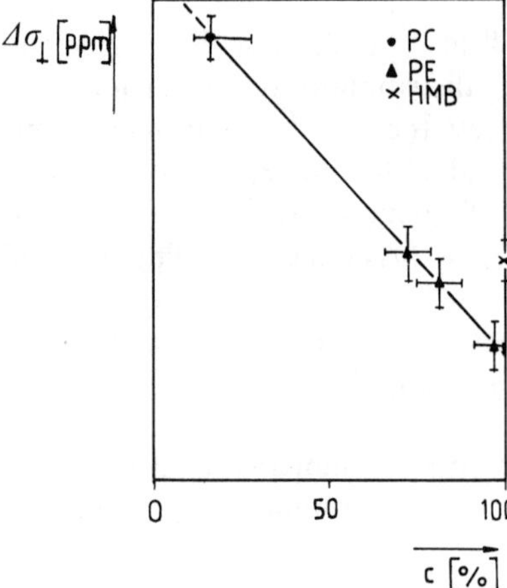

Fig. 36. Linear dependence of the width at half height $\Delta\sigma_\perp$ of the chemical shift distribution on the degree of crystallinity c of the sample

(i) From the explanation of the tensor element distributions it is clear that with the increase of the "amorphousness" the distributions become broader. Figure 36 shows an excellent linear behaviour of this width versus the degree of crystallinity (for HMB the contribution of the $^{13}C–^{13}C$ interaction has to be subtracted), which was of an unexpected accuracy of linearity. The decision, whether this surprisingly high linearity is caused by accident or not, can only be taken after further experiments.

(ii) In many cases the calculated distribution function can be divided into two or more parts. A typical example can be seen in Fig. 37, where the methyl line of polycarbonate, which is axially symmetric as a result of methyl group rotation, and

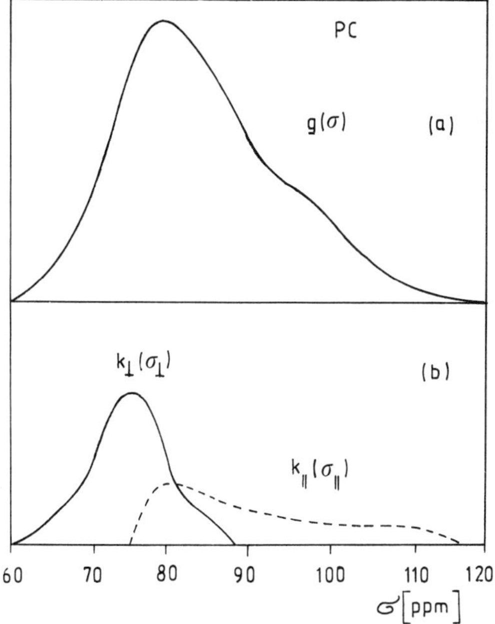

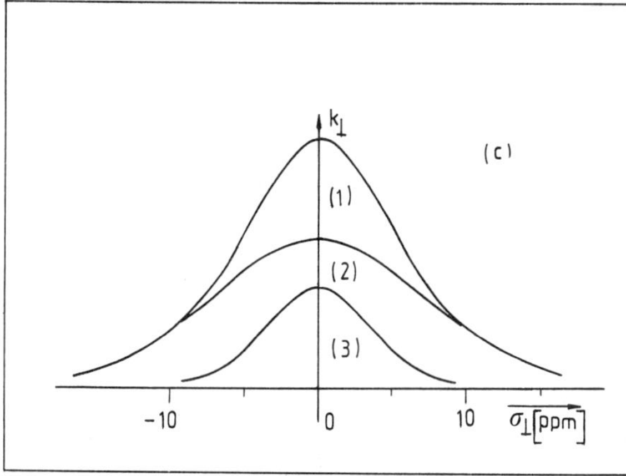

Fig. 37. ^{13}C cross-polarization spectrum (**a**) $g(\sigma)$ of the methyl-group in polycarbonate, (**b**) distribution functions $k_\perp(\sigma_\perp)$ and $k_\parallel(\sigma_\parallel)$, and (**c**) decomposition of $k_\perp(\sigma_\perp)$ (*1*) into two components (*2* and *3*)

the distribution functions $k_\perp(\sigma_\perp)$ and $k_\parallel(\sigma_\parallel)$ are represented. The decomposition of $k_\perp(\sigma_\perp)$ into two parts with Gaussian shape, for which the ratio of intensities is 3:1 and that of the widths 2:1, is also represented, showing the existence of two different regions in isotropic polycarbonate, where the smaller one is more ordered.

(iii) Eq. (144) enables us to evaluate the shielding tensor elements in dependence upon the conformation in the polymer chains. The spread of these values is greater than that caused by the variation of the intermolecular distances. For the application to polyethylene we consider ethylene pentades, where the arrangement of two pairs of carbons with respect to the centered carbon is characterized by *trans*

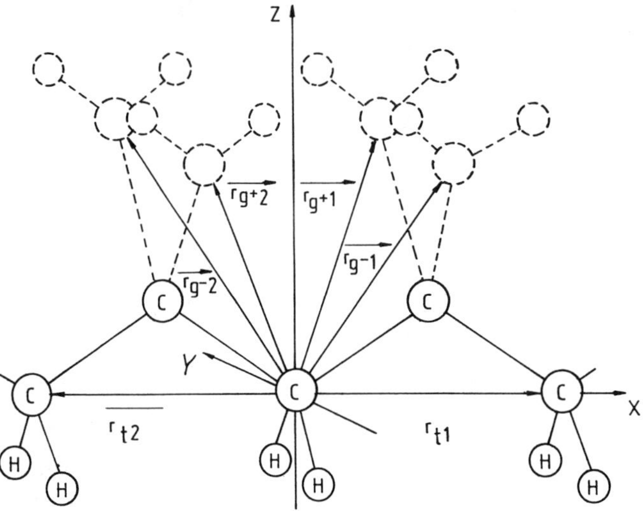

Fig. 38. Possible conformations of a methylene pentade of a polyethylene molecule

(t), *gauche* + (g^+), and *gauche* − (g^-) conformations (cf. Fig. 38). For the numerical evaluations the tensor components in the presence of trans conformations in ultra-oriented polyethylene are used and σ_0 for the bared nucleus (not exactly known from quantum mechanics) is retained in the results as a parameter for fitting the experimental results. The calculations show an increase in the shielding tensor components σ_{11} and σ_{22}, but a decrease in σ_{33} (so that the isotropic mean value remains constant) in the sequence $(g^\pm g^\pm)$, $(g^\pm g^\mp)$, $(tg^\pm, g^\pm t)$, (tt) for the nine possible conformations in the methylene pentade. If w is the probability for the *trans* conformation, then these for the *gauche* + and *gauche* −, respectively are both $(1/2)(1-w)$ and for the sequence mentioned above $(1/4)(1-w)^2$, $(1/4)(1-w)^2$, $(w/2)(1-w)$, and w^2, respectively. Comparing these results with the distribution function $k_1(\sigma_{11})$, deconvoluted as described in Sect. B using a Lorentzian function with a width of the order of the additional broadening, we found for the part that corresponds to the (tt)-signal a relative intensity of about 5/6. Hence, $w \approx 0.91$, i.e. approximately 90% of our sample have *trans*-conformations, whereas the probability for both *gauche*-conformations together is about 10%.

3.3.2 Oriented Polymers

3.3.2.1 Polymers with Weak Orientation

A) Distribution of Orientation

At the end of Sect. 1.1.2 it was shown how the distribution $f(\cos\theta)$ in anisotropic polymers of axial symmetry can be derived from the spectra of both an isotropic sample and oriented one with the draw direction parallel to the magnetic field. The assumption of axial symmetry can often be fulfilled using relatively high tem-

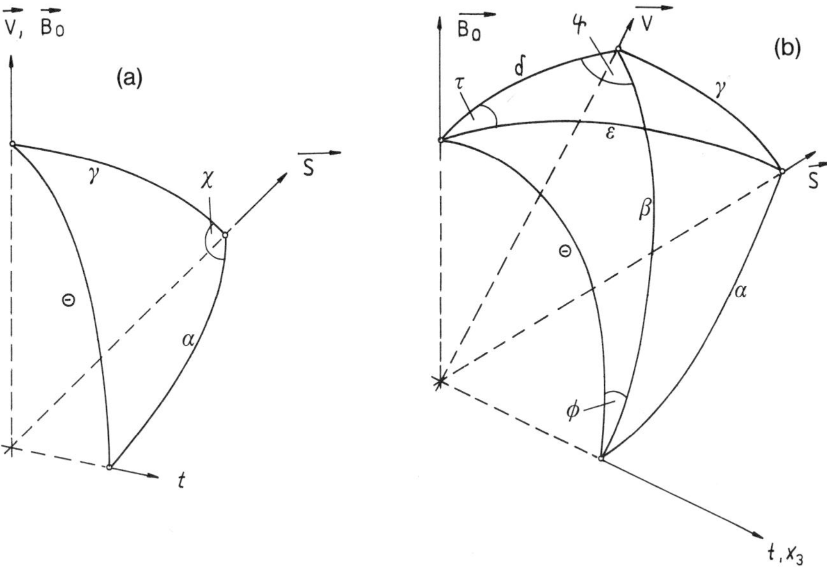

Fig. 39. Definition of the angles between the segment vector $\vec{S}$, draw-direction $\vec{V}$, static magnetic field $\vec{B}_0$, and the symmetry axis t or the axis x_3 of the shielding tensor. (**a**) $\vec{V}$ parallel to $\vec{B}_0$, (**b**) arbitrary $\vec{V}$-direction

peratures, provided that a restricted rotation about one axis causes the axial asymmetry (cf. Sect. 1.1.2). The distribution $f(\cos\theta)$, however, is not interesting in regard to the polymer structure, because θ represents the angle between the tensor symmetry axis t and the static magnetic field and is an NMR quantity even if the draw-direction $\vec{V}$ lies parallel to $\vec{B}_0$ (cf. Fig. 39a). More characteristic for the polymer structure is the distribution $k(\cos\gamma)$ of the segment orientation γ with respect to the draw direction. According to Fig. 39a the relation between the angles θ and γ is given by

$$\cos\theta = \cos\gamma\cos\alpha + \sin\gamma\sin\alpha\cos\chi. \tag{147}$$

For the angle χ, which describes the rotation of a segment around its own axis, we assume an equal probability. The θ-distribution for a fixed angle γ results from

$$|f_\gamma(\cos\theta)d(\cos\theta)| = \frac{1}{\pi}|d\chi| \tag{148}$$

and yields with Eq. (147)

$$f\gamma(\cos\theta) = \frac{1}{\pi}(1 - \cos^2\alpha - \cos^2\gamma - \cos^2\theta + 2\cos\alpha\cos\gamma\cos\theta)^{-1/2}. \tag{149}$$

Introducing the distribution $k(\cos\gamma)$ and some considerations as to the limits follows

$$f(\cos\theta) = \int_{\max[\cos(\alpha+\theta),\,0]}^{\cos(\alpha-\theta)} k(\cos\gamma)f_\gamma(\cos\theta)d(\cos\gamma), \tag{150}$$

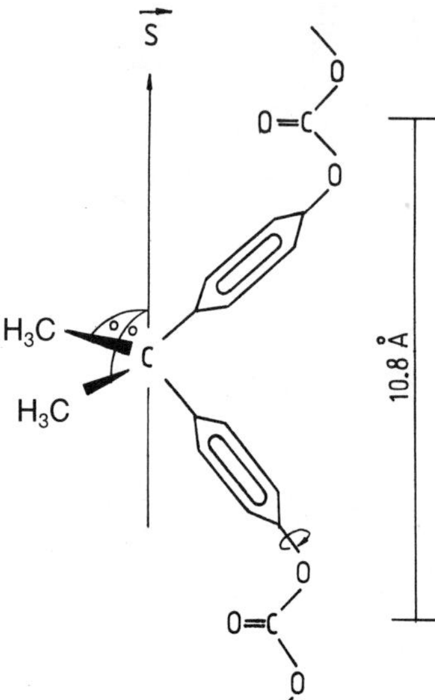

Fig. 40. Spatial structure of polycarbonate

or with the abbreviations $x = \cos\gamma$, $a = \sin\alpha \sin\theta$, $b = \cos\alpha \cos\theta$

$$f(\cos\theta) = \int_{\max[(b-a),\,0]}^{b+a} \frac{k(x)dx}{[a^2 - (x-b)^2]^{1/2}}. \tag{151}$$

This result can be simplified if the angle between the tensor symmetry axis and the segment orientation is $90°$, which e.g. is well fulfilled for the methyl groups in polycarbonate (cf. Fig. 40):

$$f(\cos\theta) = \frac{1}{\pi} \int_0^{\sin\theta} k(\cos\gamma)[\sin^2\theta - \cos^2\gamma]^{-1/2}\, d(\cos\gamma). \tag{152}$$

Replacing $\sin\theta$ by a and $\cos\gamma$ by x we get

$$f([1-a^2]^{1/2}) = \frac{1}{\pi} \int_0^{a} \frac{k(x)\,dx}{(a^2 - x^2)^{1/2}}. \tag{153}$$

Because of the equivalent feature of this integral with that in Eq. (146) (the upper limit occurs as a parameter in the integrand) after a substitution of $f(a)$ by $f([1-a^2]^{1/2})$ again a numerical integral inversion can be performed.

Another way for the determination of the orientational distribution from the chemical shift anisotropy was developed by Resing and coworkers [168]. The denotation of axes and angles is shown in Fig. 39b, where $\vec{V}$ here is an "averaged" director of the symmetry axis of the chemical shift tensor which again is assumed to be axially symmetric. For ψ a uniform distribution is proposed, but the distribution in β is given by a Rayleigh or similar distribution. Using the Eqs. (32) and (33) and substituting the angle θ by β, ψ, and δ according to

$$\cos\theta = \cos\beta \cos\delta + \sin\beta \sin\delta \cos\psi \tag{154}$$

the formulae for the observed line shapes can be derived in an analogous manner as for the determination of $f(\cos\theta)$ in Eq. (150). Using the distribution function $p(\cos\beta)$ instead of $p(\beta)$ in order to get an equal distribution function for isotropic samples and to make the estimation of the orientational conditions from the distributions easier, in the limiting case if the director is parallel to the magnetic field ($\delta = 0$, i.e. $\beta = \theta$) one of course directly gets Eq. (33). In [169, 170, 15] the general equation for arbitrary angles δ of the director with respect to the magnetic field is presented quite equivalent to Eq. (150):

$$G_\delta(\sigma) = I(\sigma)f(\cos\theta), \tag{155}$$

where in $f(\cos\theta)$ from Eqs. (150) or (151) the angles α, γ as well as the distribution $k(\cos\gamma)$ are to be replaced by δ, β, and $p(\cos\beta)$, respectively, and $\cos\theta$ is given by Eq. (34). The task to interpret a measured spectrum consists in its fitting using Eq. (155) with a suitably chosen distribution function $p(\cos\beta)$ and in the determination of the parameters. It can be shown that in comparison with the limiting cases $\delta = 0$ and $\pi/2$ the greatest effect of the distribution $p(\cos\beta)$ of the orientation results for $\delta = \pi/4$ giving the broadest line. This method represents an extension in relation to that discussed before and yield more experimental possibilities, because the former contains the case $\delta = 0$ only. However, two other disadvantages have to be removed by techniques used in the former calculation in order to extend the applicability:

(i) The choice of the analytic expression for $p(\cos\beta)$ means an *a priori* restriction. Principally a formalism similar to that applied for Eqs. (146) and (153) is possible for Eq. (155) as well. In this way $p(\cos\beta)$ can be determined numerically step by step from the quotient $G_\delta(\sigma)/I(\sigma)$.

(ii) If the tensor symmetry axis does not agree with the segment orientation in the polymer chain, the distribution of the former cannot be used for direct interpretation (cf. e.g. $f(\cos\theta)$ in comparison with $k(\cos\gamma)$ in Fig. 42). The transformation into a distribution of the segment orientation with respect to the director similar to the treatment mentioned above has been given by Eq. (150), where according to Fig. 39 θ has to be replaced by β. Therefore, from the distribution $p(\cos\beta)$ calculated by means of Eq. (155) the determination of the distribution $k(\cos\gamma)$ is possible which describes the orientational relations of the segment directions with respect to the director or the draw direction, respectively. Taking into account all these recommendations a powerful method arises for the determination of orientational distributions in weakly oriented polymers.

A general discussion of the possibilities to get the distribution function $k(\cos\gamma)$ for the chain segments with respect to the direction of orientation is carried out by Spiess [171] (cf. [172, 173] also). Because of the highly adequate procedure for several anisotropic couplings, the frequency of a single NMR transition depend upon the orientation of the magnetic field with respect to the frame of the principal axes of the respective coupling in the same manner as the chemical shielding $\sigma - \sigma_{iso}$ in Eq. (17):

$$\omega = \delta[(3\cos^2\theta - 1) + \eta\sin^2\theta\cos 2\phi]. \tag{156}$$

The meaning of δ for the anisotropic chemical shift, and for the dipole–dipole interaction both for like and unlike spins as well as for quadrupolar interaction is represented in a table in [171].

Particularly for the former interaction, δ is equivalent to $1/2\gamma B_0\delta'$ with δ' according to Eq. (15) and for the latter it corresponds with δ from Eq. (39). For the quadrupolar coupling however, for which in [171] most applications are given, this treatment is restricted to nuclei with small quadrupolar interactions, e.g. deuterium, in order to enable the application of the first order perturbation theory. Spiess demonstrates three ways for the calculation of the distribution function $k(\cos\gamma)$ in order to represent the NMR line shape as a superposition of a relatively small number of subspectra, which can easily be calculated analytically. The first method is an expansion of the orientational distribution in terms of Legendre polynomials; it is restricted, like the methods mentioned above, to axial symmetry and particularly useful for analyzing the spectra of moderately ordered samples. Fitting the calculated line shape to an experimental spectrum, the moments of the orientational distribution can be determined. Here also an experimental variation of the angle δ (cf. Fig. 39b) of the static magnetic field with respect to the direction of orientation can yield favourable conditions for interpretation.

The second method consists in an expansion in terms of planar-, the third in terms of conical distributions. Starting point in both cases is an ensemble of parallel chains with transverse isotropy, i.e. no order exists in the plane perpendicular to the chain axis. These completely ordered subsystems cause line shapes with two or three singularities in dependence on the angle ε between the common segmental direction and the magnetic field. This common molecular axis lies for the planar distribution in the direction of one of the principal axes of the coupling tensor, whereas the other two axes are in parallel planes for all nuclei of the ensemble. For such a distribution it is possible to calculate the line shape even if the coupling tensor lacks axial symmetry. If, for example, the principal axis X_1, X_2 or X_3, respectively, is the common one, the line shape for an ensemble characterized by an angle ε, is

$$G_\varepsilon(\omega) = \pi^{-1}(A^2 - B^2)^{-1/2} \tag{157}$$

for $|B| < A$, where

$$A = a\delta\sin^2\varepsilon, \quad B = \omega - b\delta(3\cos^2\varepsilon - 1), \tag{158}$$

$$a = \tfrac{1}{2}(3 - \eta), \tfrac{1}{2}(3 + \eta) \quad \text{or} \quad \eta \tag{159}$$

and

$$b = -\tfrac{1}{2}(1 + \eta), -\tfrac{1}{2}(1 - \eta) \quad \text{or} \quad 1, \tag{160}$$

respectively.

If the common molecular axis is not parallel to one of the principal axes, one assumes a uniform distribution on a cone of the principal axes for a given ensemble, forming an angle α with the common direction. The line shape for such an ensemble then depends again on the angle ε and also parametrically on α and is given in [171] for axial symmetry ($\eta = 0$). For the determination of the line shape of partially ordered polymers of course a second step is necessary using the latter two methods. One has to produce a superposition of the line shapes $G_\varepsilon(\omega)$ for the planar or conical distribution with the weighting factor $Q_\delta(\varepsilon)$, which also depends on the experimental choice of the angle δ getting

$$G(\omega) = \tfrac{1}{2} \int\limits_{-1}^{+1} Q_\delta(\varepsilon) G_\varepsilon(\omega)\, d(\cos \varepsilon). \tag{161}$$

Finally $Q_\delta(\varepsilon)$ has to be calculated from the only interesting distribution function $k(\cos\gamma)$ by variation of the angle τ (cf. Fig. 39b) according to

$$Q_d(\varepsilon) = \int\limits_{-\pi}^{+\pi} k(\cos\gamma)\, d\tau, \tag{162}$$

where γ is related to δ, ε and τ by

$$\cos\gamma = \cos\varepsilon \cos\delta + \sin\varepsilon \sin\delta \cos\tau. \tag{163}$$

The numerical calculation of the line shape can be performed using Eqs. (157–163) or the respective expression for the conical distribution [171] instead of Eq. (157) for any $k(\cos\gamma)$. Contrarily to the methods proposed by Hempel and Resing for planar distributions (and also for conical ones with numerically evaluated shapes for $G_\varepsilon(\omega)$) the calculation is also possible for systems lacking axial symmetry. Moreover, the planar and the conical distributions are also well suitable for highly oriented polymers (cf. Sect. 3.3.2.2), because the subspectra belong to completely ordered ensembles of molecules. However, it turns out to be a disadvantage that a proper choice of an analytic form $k(\cos\gamma)$ is necessary and the parameters are determined by fitting the experimental spectrum, whereas the previously mentioned methods can calculate the distribution $k(\cos\gamma)$ directly from the experimental line shape using the integral inversion method which is not applicable for the latter method.

B) Applications

The methods were applied for polycarbonate. Both spectra, for isotropic samples and materials stretched by a factor of 1.7, are shown in Fig. 41.

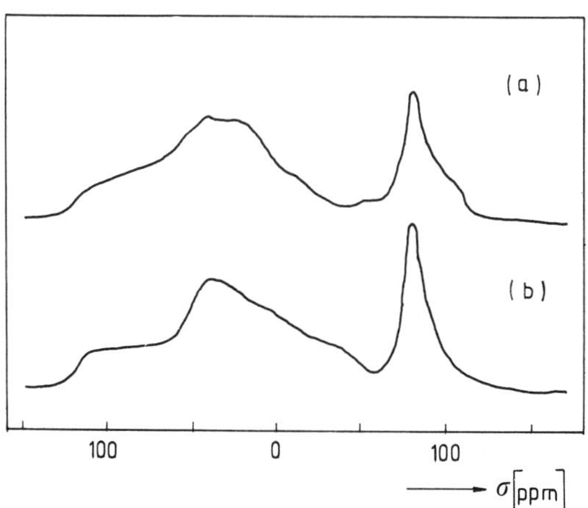

Fig. 41. ^{13}C cross-polarization spectra of polycarbonate of (a) isotropic and (b) stretched material. Right signal: CH$_3$-groups, left broad line: all other carbons

The distribution functions $f(\cos\theta)$ and $k(\cos\gamma)$, calculated for the methyl line using Eqs. (33) and (153) are depicted in Fig. 42, the latter in Cartesian as well as in polar coordinates. From these representations we establish a heterogeneity of the sample consisting of an isotropic (about 65%) and of a relatively highly oriented part (about 35%). The latter only contains orientations of the segments with respect to the draw-direction smaller than 60° with a maximum for $\gamma = 0°$. This could be interpreted as follows: The stretched sample consists of two regions without and with oriented chain segments. In comparison with the "degree of orientation", which is often determined by the use of X-ray small angle diffraction and which represents a more general integral parameter, the knowledge of the γ-distribution contains much more detailed information concerning the distribution of segment

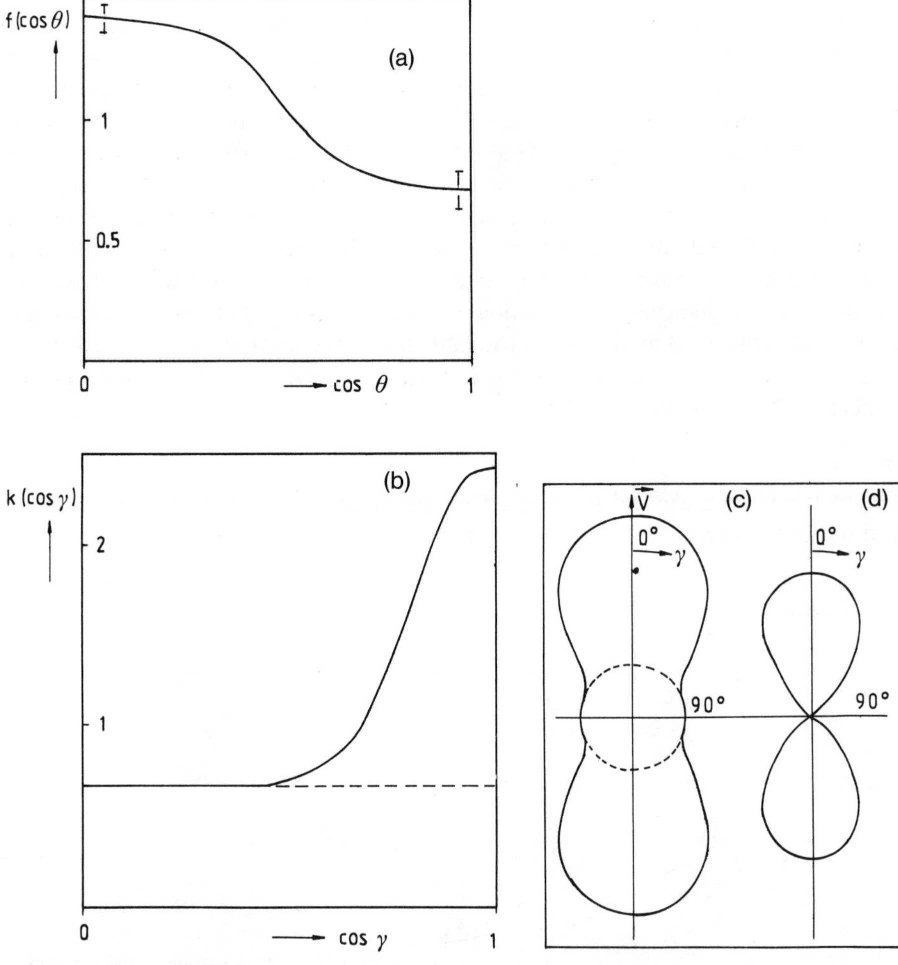

Fig. 42. Distribution functions for the orientation of the methyl group in polycarbonate: (a) $f(\cos\theta)$, (b) $k(\cos\gamma)$ in Cartesian coordinates, (c) $k(\cos\gamma)$ in polar coordinates, and (d) as (c), however the isotropic part is subtracted

orientations. Besides an estimation of regions with different orientations, the probability of the abundance of segment orientations in a defined angle interval can be read. Nevertheless, we also derive from distribution $k(\cos \gamma)$ two integral parameters which are mean values and characterize the orientation degree of the sample. They are equal to zero in the isotropic and equal to one in the fully oriented case:

$$S_1 = 2\,\overline{\cos\gamma} - 1, \tag{164a}$$

$$S_2 = \tfrac{1}{2}(3\,\overline{\cos^2\gamma} - 1). \tag{164b}$$

As expected from the experimental results, for the isotropic part follows $S_1 = S_2 = 0$, whereas for the highly oriented region $S_1 = .74$, $S_2 = .64$ and for the whole sample $S_1 = .26$, $S_2 = .23$. The latters are in good agreement with the degree of orientation of about 25% mentioned above and determined by X-ray measurements. At last we want to apply some statistical treatment for the interpretation of the distribution function $k(\cos \gamma)$. Taking into consideration a number of 400 segments per molecule and a segment length of 1.08 nm, one gets for the mean length $L_{\parallel}$ and width $L_{\perp}$ of a polycarbonate chain parallel and perpendicular with respect to the draw direction in the highly ordered region [15, 169] $L_{\parallel} = 374$ nm and $L_{\perp} = 10.5$ nm.

Although the sample is stretched only by a factor of 1.7, surprisingly the mean ratio between length and width of a polymer chain is 36. This cannot be understood without further discussion. We remember that, as was discussed in Sect. 3.3.1, even in isotropic unstretched samples of polycarbonate two regions with different ordering states do exist. However, the orientations of these small subunits are statistically distributed and therefore they do not cause a macroscopic orientation. The stretching process then turns these "microcrystals" in the draw-direction.

The application of the deuteron NMR for the determination of the orientational distribution is demonstrated in [171] and other papers by Spiess and Sillescu quoted herein. For linear polyethylene a single crystal and a sample drawn by a factor $\lambda \approx 9$ were investigated. In the single crystal the line shape investigations can be explained under variation of the angle δ of the direction of orientation with respect to the magnetic field by an expansion of the orientational distribution in terms of the first eight Legendre polynomials. This results in a Gaussian curve with a dispersion of $\bar{\gamma} \approx 12°$ and an additional isotropic contribution corresponding to 20% of the total intensity. The drawn sample shows a more drastic dependence on the angle γ and can be interpreted by fitting the spectra using the expansion in terms of planar distributions and again assuming a Gaussian for $k(\cos \gamma)$ with a dispersion $\bar{\gamma} \approx 3°$. The agreement with X-ray results is quite good. Also for the characterization of the degree of order in the amorphous regions of semicrystalline drawn polyethylene the ^{2}H NMR was applied [171]. Measurements were performed at 133 K in order to slow down the molecular motion so that it did not average out the quadrupolar coupling. Separating the signal of the amorphous regions by utilizing the T_1-values shorter by about one order compared with those in the crystalline regions (cf. Sect. 3.4.1.1) one gets of course a lower degree of order in the amorphous regions (characterized by a width of the orientational distribution of approximately $\pm\,12°$) as compared with the crystalline regions. A more detailed discussion leads to the conclusion that an appreciable number of CH-bonds must form angles with respect to the draw direction in the vicinity of $35°$ corresponding to the *gauche* conformations present in the amorphous regions.

3.3.2.2 Highly Oriented Polymers

A) Orientation and Signal Position

In fibres or in polymers highly oriented by other means the aim is to determine an averaged value for the inclination γ between the orientation of the segments $\vec{S}$ and the draw direction $\vec{V}$ using the angle β between the draw direction and the principal axis relating to the eigenvalue σ_{33} (cf. Fig. 39b).

The angle δ between the draw direction and the static magnetic field $\vec{B}_0$ is important for the investigator. Substituting θ and ϕ in Eq. (13) by δ, β, and ψ and assuming an equal distribution for ψ we get the line shape $G(\sigma)$ in dependence on δ and β in a similar manner as in Sect. 1.1.2. The complicated result derived from [15] has the form

$$G(\sigma) \sim \frac{1}{C^{1/2}} \{ [A^2 - (B + C^{1/2})^2]^{-1/2} + [A^2 - (B - C^{1/2})^2]^{-1/2} \}, \quad (165)$$

where

$$A = D \sin \delta \tag{166a}$$

$$B = (\sigma_{33} - \sigma_{11}) \sin \beta \cos \beta \cos \delta, \tag{166b}$$

$$C = (\sigma - \sigma_{22})D + (\sigma_{33} - \sigma_{22})(\sigma_{22} - \sigma_{11}) \cos^2 \delta, \tag{166c}$$

$$D = \sigma_{11} \cos^2 \beta - \sigma_{22} + \sigma_{33} \sin^2 \beta. \tag{166d}$$

$G(\sigma)$ describes the line shape within the three singularities (resulting from setting each of the roots equal to zero) and vanishes outside those.

Except for the trivial case, when the draw-direction is parallel to the magnetic field $\vec{B}_0 (\delta = 0)$ and all nuclei have the same chemical shift producing a narrow single line at the position

$$\sigma_s = \sigma_{11} \sin^2 \beta + \sigma_{33} \cos^2 \beta \tag{167}$$

(which follows formally from setting $C = B^2$), all other positions of the singularities for arbitrary angles β and δ can be determined. In the cases of practical interest ($B = 0$, i.e. $\delta = 90°$, $\beta = 0°$, or $\beta = 90°$, respectively) one gets two singularities at $C = 0$ and $C = A^2$. Especially for $\delta = 90°$ if the draw direction is aligned perpendicular to $\vec{B}_0$, one observes a spectrum with two peaks at

$$\sigma_{D1} = \sigma_{22}, \tag{168a}$$

$$\sigma_{D2} = \sigma_{11} \cos^2 \beta + \sigma_{33} \sin^2 \beta. \tag{168b}$$

These special results agree, of course, with others published earlier (e.g. [174]), where an application is also shown for ultraoriented polyethylene for which $\beta = 0$.

B) Applications

Let us consider an example described in [13]. Poly-p-phenylene terephthalamide is a highly oriented fibre the structure of which is shown in Fig. 43, the spectra for the isotropic and the stretched sample with the draw direction parallel and perpendicular with respect to the magnetic field $\vec{B}_0$ are represented in Fig. 44.

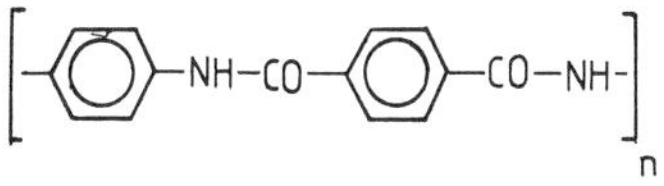

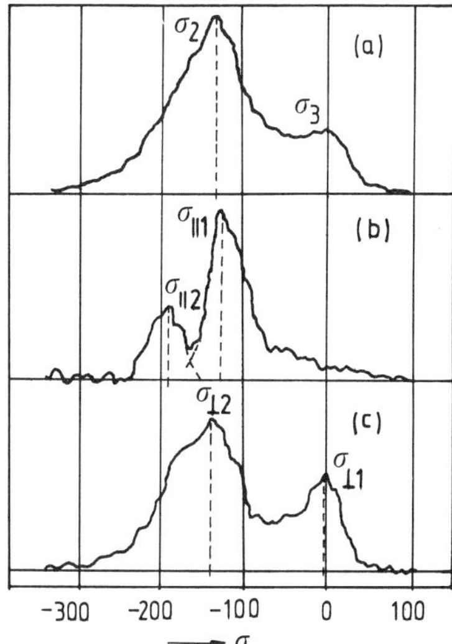

Fig. 43. Structure of poly-p-phenylene terephthalamide

Fig. 44. ^{13}C cross-polarization spectra of terlon of isotropic (a) and stretched material with the fibre axis parallel (b) and perpendicular (c) with respect to the $\vec{B}_0$ field

The considerable broadening of the carbons directly in the neighbourhood of the ^{14}N nuclei is here an advantage, because their signals do not contribute to the spectrum and there are only three kinds of chemically equivalent ^{13}C nuclei with nearly the same shielding tensor elements following from the isotropic spectrum:

$$\sigma_{33} = -195 \pm 10 \text{ ppm} \quad \text{in the direction of the C–H-bonding,}$$
$$\sigma_{22} = -135 \pm 5 \text{ ppm} \quad \text{tangential to the ring,}$$
$$\sigma_{11} = 0 \pm 10 \text{ ppm} \quad \text{(for reference) perpendicular to the ring plane.}$$

Even a superficial consideration of the three spectra makes it possible to conclude that the draw direction lies in the plane of the phenyl rings. For an assumed angle ε between the draw direction $\vec{V}$ and the ring plane, ε appears also as the angle between $\vec{V}$ and the principal axis X_3, when one considers the ring carbons neighbouring the carboxylic groups. In the parallel-spectrum these ^{13}C-nuclei lead to a line at a position given by Eq. (167) with $\beta = \varepsilon$. Comparison with the experiment only shows signals in the range from σ_{22} to σ_{33}, i.e. ε lies between $0°$ and $35°$.

On the other hand, in the perpendicular-spectrum the carbons mentioned above produce a doublet with positions given in Eq. (168) again with $\beta = \varepsilon$. The peak at σ_{22} is well visible, the other lies at $\sigma_{11}(\varepsilon = 0)$ or between σ_{22} and σ_{33}, yielding ε-values from ca. 55° up to 90°. The only common explanation is an angle of $\varepsilon = 0°$, q.e.d. We will not continue the interpretation in such detail. On the basis of the position of the fibre axis in the phenyl-ring plane one can propose a few possible arrangements for the terlon chains and discuss the expected spectra using the equations derived above and considering now, of course, all ^{13}C-nuclei contributing to the spectra. Again comparing with the experimental spectra one has to decide (possibly, as in the latter example, by introduction of a distribution of arrangements between two limiting cases) what kind of arrangement is in best agreement with the experimental data, taking into account a smoothing of the signals as a result of several interactions (cf. Sect. 3.3.1), particularly in the case of the strong dipolar coupling with the ^{14}N-nuclei.

3.3.3 Further Methods for the Determination of the Polymer Structure

3.3.3.1 Observation of Dipolar ^{13}C–^{13}C Satellites

Using a normal cross-polarization procedure without magic angle spinning, i.e., observing spectra with full chemical shift anisotropy under the conditions of a very high signal-noise ratio and of well resolved lines, dipolar satellites have been observed in the spectra of ultraoriented polyethylene [175]. This splitting is caused by the interaction of ^{13}C-pairs, occurring very seldom because of the low natural ^{13}C abundance of 1.1%, and can be expressed by

$$\Delta\omega = \frac{3}{2}\frac{\gamma_C^2 \hbar}{r^3}(3\cos^2\theta - 1)\left(\frac{\mu_0}{4\pi}\right) \tag{169}$$

with the commonly used notations. It is quite clear that in isotropic samples such satellites cannot be observed because of the broad distribution of angles θ of the ^{13}C–^{13}C internuclear vector with respect to the magnetic field. However, if the samples are well oriented and the carbons in the backbone have a uniform geometric relationship, the satellites are smaller and observable. This connection is treated by Vander Hart [175] and makes it possible to calculate the deviation of the chains from the draw direction directly from the satellite line widths.

3.3.3.2 Separated Local Field Spectroscopy

From the second chapter, it could be understood how one can obtain information on the distances and angles between the ^{13}C-nuclei and the neighbouring protons by evaluation of the dipolar split spectra for fixed values of chemical shift in a two-dimensional plot. Here, however, we shall consider the first applications of the SLF-method in polymers [124, 176] for highly oriented polyethylene, where a single line, e.g. an anisotropic powder pattern occurs instead of different lines for several chemically nonequivalent carbons. The main chains lie in draw direction which agrees with that of the x_3-shielding tensor axis. A sharp single line at the position σ_{33} follows from Eq. (167) with $\beta = 0$ when the orientation is parallel to the $\vec{B}_0$-field ($\delta = 0$). With the orientation perpendicular to the field ($\delta = 90°$) a spectrum with singularities at σ_{11} and σ_{22}, respectively, is obtained. The interaction with the two equivalent protons gives rise to a dipolar splitting in a triplet with the intensity ratios $\approx 1:2:1$ and a distance between two lines according to

$$\Delta\omega = \frac{\gamma_H \gamma_C \hbar}{r_{CH}^3}(3\cos^2\theta_{CH} - 1)\left(\frac{\mu_0}{4\pi}\right). \tag{170}$$

For the single line the angle θ_{CH} of the C–H-bonds with the magnetic field parallel to the segment orientation is always $90°$. The expected triplet with intensity ratios mentioned above and an overall splitting $2 \cdot \Delta\omega/2\pi$ of 45 kHz can be explained with a bond length $r_{CH} = .11$ nm. If the chain orientation is perpendicular to the magnetic field, we define the other two principal axes for the discussion of the powder pattern: x_1 normal to the C–C–C plane, x_2 along the axis bisecting the H–C–H angle. Therefore the angles θ_{CH} following from the singularities σ_{11} and σ_{22} are given replacing the H–C–H bond angle χ by $90° - \chi/2$ and $\chi/2$, respectively. The experimental fact that only an unsplit line occurs at one position σ_{11} or σ_{22} in the "dipolar direction" means that the proper angle θ_{CH} is the magic one. The more realistic of the two possibilities is $\chi = 109.5°$ according to σ_{22}. The remaining dipolar triplet corresponds to σ_{11} and to the angle $\theta_{CH} = 90° - \chi/2$ and again makes it possible to determine the bound length. Besides imperfections in the experiment and data processing this also yields the predicted results.

Of course a generalization of this treatment for more complicated polymers (cf. e.g. [177]) and for isotropic samples would be connected with new difficulties. However, the gain of detailed structural information justifies the high expense of this two-dimensional method.

3.3.3.3   Rotor-Synchronized Two-Dimensional MAS Spectra

A rotor-synchronized two-dimensional magic-angle spinning technique described roughly in Sect. 2.2.3 was used [281, 282] for the determination of structure and order in partially oriented polymers using ^{13}C-sideband spectra. The experiments have been applied to systems with a preferential axis of orientation not parallel to the rotor axis. They allow one, as demonstrated for highly oriented PE [281], to detect the presence of such a preferential axis and its relationship with the rotor axis as well as with the chemical-shielding tensor principal-axis system. Moreover, the degree of orientation can be expressed as a distribution of orientations of the director. This distribution function can be obtained [282] even for the crystalline and amorphous regions in semicrystalline PETP.

The main advantage of this two-dimensional technique consists in a resolving of different residues of a molecule in one dimension and in the determination of the degree of order separately for these residues in the other one. Therefore, this method can be very useful for structural studies of oriented synthetics and biopolymers.

3.3.4 Methods for the Reestablishment of Anisotropy Parameters in Polymers

In the previous sections it was shown that the anisotropy properties particularly in polymers lead to a good deal of important structural information. This, however, was demonstrated only for special examples. The reason is that the spectra in polymers are very complicated (cf. Fig. 45) [178], and seldom is it possible to separate several powder patterns for special chemically equivalent carbons. To characterize the anisotropy the position of the singularities of the spectra, the ratios of the intensities or of the area parts, the moments of the lines or the line shapes themselves (cf. Chap. 1) can be used. On the other hand, fast sample spinning at the magic angle, necessary for resolution enhancement, unfortunately, eliminates the anisotropy of the spectra. Therefore, for a wider application of the methods described in this section a restoration of the anisotropy properties without loss of the higher resolution is required.

Now follows a short discussion of two possibilities for the realization of this task.

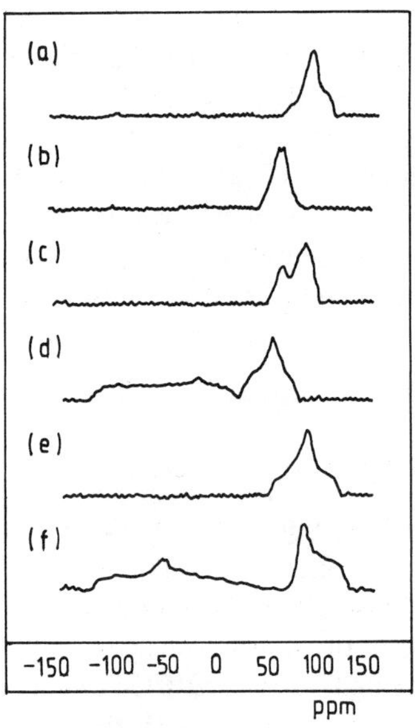

Fig. 45. Cross-polarization spectra of some polymers. (**a**) low-density polyethylene, (**b**) poly(vinylchloride), (**c**) poly(vinyl alcohol), (**d**) poly(vinyl acetate), (**e**) poly(ethylene vinylacetate), (**f**) poly(carbonate)

3.3.4.1 Deviations of the Magic Angle

Considering Eq. (40) it is quite clear that for deviations from the magic angle the anisotropy will be reduced by the factor $\frac{1}{2}(3\cos^2\beta - 1)$. Experiments verify this influence of the angle β of the spinning axis with respect to the static magnetic field and even the change of the sign of the anisotropy by passing the magic angle [179–181]. That means that for a suitable angle the lines can often be separated from one another, but the anisotropies are still measurable. However, the reduction of the anisotropy makes its evaluation more difficult and, moreover, the spinning angle has to be measured exactly. This approach is especially useful when the isotropic values of the lines are well separated, as it is e.g. in the case of the signal from the carboxylic carbon in poly-(methyl methacrylate).

3.3.4.2 Slow Spinning at the Magic Angle

In high-field super-conducting solenoids slow spinning at the magic angle (i.e. the spinning frequency ω_r is slower than the anisotropy of chemical shift but higher than the remaining linewidth under strong dipolar decoupling and rapid magic angle spinning conditions) is often a necessity because of the limitation of the mechanical spinning frequencies ω_r. It gives rise to rotational spin echoes in the free induction decay and, correspondingly, to sidebands in the frequency spectrum. The Fourier transform of the echo envelope is the isotropic spectrum, whereas that of the echo shape and accordingly the intensity distribution of the sidebands contains information concerning the anisotropies. Using this knowledge some treatments has been developed, which allow the determination of anisotropy parameters by a moment analysis (applying Eqs. (27) and the equivalent formulae for the higher moments, which are functions of $\langle \Delta\omega^2 \rangle$, $\langle \Delta\omega^3 \rangle$, and ω_r, respectively [182]), from the intensities of only few spinning side bands [183–185] and by the use of rotational-synchronized *rf* π-pulses in the ^{13}C-channel after each half or quarter cycle. These additional pulses during the evolution time (cf. Table 3) allow the development of the spin system under the influence of the anisotropic part of the chemical shift Hamiltonian. Therefore, variation of the evolution time and two-dimensional Fourier transformation permit the determination of chemical shift anisotropy without any reduction in the spectral resolution. Although the experimental realization of the latter method is very complicated, it turns out to be an advantage in comparison with the other methods, if not only the anisotropy parameters themselves but also the line shape is to be investigated.

Examples for the application of rotational spin echoes in solids are given in [186] for poly(*n*-butyl methacrylate) and of the sideband technique in [187] for poly(butylene terephthalate) and related copolymers.

3.4 Improvement of the Structural Information by Different Techniques

3.4.1 Suppression of Parts of the Spectrum by the Use of the Differences in the Relaxation Times

To interpret the line shape (cf. Sect. 3.3) and to determine chemical shifts in highly resolved isotropic spectra (cf. Sect. 3.5) or to measure the relaxation times themselves (cf. Chap. 4) it is often favourable and even necessary to enhance or to suppress parts of the spectrum. This can be done by means of the fact that those parts of the spectrum which are e.g. characterized by different chemical structural units, conformations, molecular orders, oscillations and mobilities have different relaxation times. Several variants should be presented utilizing different relaxation times for the discrimination of different parts in the spectrum. Another successful way for the enhancement of interesting spectral regions by selective enrichment of ^{13}C or ^{2}H will be discussed in Sect. 3.4.3.

3.4.1.1 Differences in T_1 values

A) Basic Experiment

Taking into account the different T_1 values for several carbons one can perform a simple $\pi/2$ (^{13}C) experiment followed by data acquisition with dipolar decoupling of the protons (cf. remarks on Table 3). For those carbons with relaxation times greater than the experiment repetition time t_r the spectra are not fully relaxed and the magnitude of the signal is strongly damped by a factor $[1 - \exp(-t_r/T_1)]$. (This effect is also used for the T_1 measurement in the progressive saturation method, cf. Table 2.) Hence, one can observe the spectra of carbons with shorter relaxation times separately. Considering the aromatic part, well separated from the aliphatic one in a polyester, in [189] a single $\pi/2$ (^{13}C)-pulse excitation with a short repetition time of 1 s is applied leading to the suppression of the signal of the nonprotonated aromatic and carbonyl carbons because of their long $T_1 - {}^{13}$C values and allowing the increase of the anisotropic spectrum of the other aromatic carbons which are all chemically equivalent.

The same experiment was performed in [190] (cf. [191] also) and clearly shows for semicrystalline polyethylene that the part of the spectrum due to the crystalline regions is more attenuated due to a very long T_1 value, when the repetition time is shortened from 10 s to 1 s.

B) Modified Experiment

A method provided by Van der Hart [192, 14] also utilize different T_1 values.

In principle this experiment can be described in the manner used for Table 2 by $[\pi/2, \tau, \pi/2, t_r/2, \pi, t_r/2, \pi/2, \text{FID}]_n$ in the ^{13}C-channel, where the repetition time t_r is 20 s, the two $\pi/2$-pulses separated by $\tau = 150 \, \mu$s ensure ^{13}C saturation, and additionally in the ^{1}H-channel during the data acquisition and after the π-pulse 30 ms-pulses are applied for dipolar proton decoupling and minimization of the transient Overhauser effect, respectively. The damping factor for the signals with different T_1 values here is given by $[1 - \exp(-t_r/2T_1)]^2$. Choosing for $t_r = 20$ s one gets for signals with $T_1 = 100$ s a contribution of about 1% in comparison with approximately 18% for method A. On the contrary, the signals with relaxation times $T_1 = 2$ s are in both cases in fact undamped (method A: 100%, B: 98.7%). That means the latter method reflects the distinction of the relaxation times more clearly.

In [192] this procedure was performed for the investigation of the non-crystalline region of drawn linear polyethylene. In this material at temperatures of 28 °C the T_1 value for carbons in a purely crystalline environment is of the order of 700 s, whereas a substantial number of carbons have $T_1 < 2$ s. The latter condition in [192] is used for a classification of those regions as noncrystalline components

(NCC). This definition and the choice of 20 s for t_r is, of course, arbitrary because of approx. 20% of carbons with T_1 values between 2 s and 700 s. However, this arbitrariness is based on the existence of an intermediate region, shown by many other methods, e.g. by proton wideline NMR. Using a simple $\pi/2$-pulse on the ^{13}C-nuclei with dipolar decoupling during the data acquisition for the definition of the total intensity, where very long repetition times of 5000 s (!) are necessary for fully relaxing all carbons, the noncrystalline fraction obtained from the NCC-spectra of drawn linear polyethylene is measured as a function of the annealing times and temperatures. Here we shall only mention one interesting aspect of those results.

The spectra of the drawn sample were obtained with the draw direction $\vec{V}$ parallel to the magnetic field $\vec{B}_0$. Therefore, the application of Eq. (167) is possible to explain the portion of the signals in different regions by corresponding angles β. Here β is the angle between the main chain and the draw direction because the principal axis x_3 lies in chain direction (cf. Sect. 3.3.2.2 or [15], where the values of σ_{11} and σ_{33} are also given, which are required for the treatment discussed here).

In this way one can establish that there is a substantial fraction of the NCC which is oriented within $15°$ of the drawing direction. After annealing these highly oriented chain segments either do not change or increase their peak intensity somewhat.

A possible but not necessary conclusion is that annealing stiffens only those chains which have been oriented by the drawing process.

C) ^{2}H Measurements

For ^{2}H measurements T_1 differences are also used for the suppression of parts of the spectrum. A pulse sequence consisting of some $\pi/2$-pulses separated by ≈ 2 ms was used a time τ_0 before the sequences for the solid– or the spin–alignment echo, respectively [277, 268]. This saturation-recovery technique enables the measurements of the T_1 and T_{1Q} relaxation using τ_0 variation, and after decomposing the signals in exponentials the determination of the amorphous and the crystalline components, e.g. in partially crystalline PE [277]. Therefore, it should be possible choosing a τ_0 value of about $5 \times T_{1a}$, however, small compared with T_{1c}, to get partially relaxed spectra containing the signal of the mobile fraction only. In comparison with the fully relaxed spectrum ($\tau_0 \approx 5 \times T_{1c}$) one can get selective information on both phases. The difference of the fully and the partially relaxed spectra gives a "Pake spectrum" [277] from which the crystallinity of PE was determined in dependence on the temperature in good accordance with Raman and X-ray data.

Moreover, the discussion of the solid–echo line shape only for the mobile deuterons yielded an interesting picture of the noncrystalline regions in PE. For different temperatures the fractions of flexible units consisting of 3, 5 or 7 bonds, respectively, and of rigid units could be estimated.

3.4.1.2 Differences in T_2-Values

In Table 3 for the usual cross-polarization experiment and for the T_1 measurement an additional waiting period τ without any irradiation is introduced before the

the data acquisition starts. That means in this short time of several tens of microseconds the proton decoupling is interrupted and a dephasing of signals ensues caused by T_2-effects which are determined by the strength of the carbon coupling to the protons. Consequently, the carbons strongly coupled to protons dephase, whereas the nonprotonated carbons retain their phase coherence during this period. Hence, the protonated carbons are suppressed in the signal as it is shown in [193] for several proteins. Of course, molecular motions, e.g. in methyl groups, can also average the dipolar coupling of carbons with protons and increase the T_2 value. Therefore, signals of methyl carbons, although protonated, are not suppressed by the technique described above. Furthermore, a separation of the crystalline and amorphous resonances in semicrystalline polymers is possible. Because of the restricted molecular motion and a greater packing efficiency the crystalline carbons are more tightly coupled to the protons leading to shorter T_2-values.

This is demonstrated in [164, 190] for the spectra and in [182, 190, 194] for the T_1-measurements of semicrystalline polyethylene, where in [190] both effects for the suppression of the crystalline component are even coupled: short repetition times via T_1 (cf. Sect. 3.4.1.1, procedure A) and interruption of proton decoupling via T_2 as described in this section.

T_2 effects can also be used in the ^{2}H solid-echo technique. If the pulse distance is large compared with T_2 of the more mobile phase but smaller than T_2 of the crystalline deuterons, the echo represents the crystalline signal only [17].

3.4.1.3 Differences in T_{CH}-Values

Differences in the cross-relaxation time CH can also be used to distinguish signals of various types of carbons. Application of Eq. (50) enables one to determine the T_{CH}-values for different carbons quantitatively [50]. Taking into account the shorter T_{CH} for the methylene carbons occurring in many polymers, a shortening of the cross-polarization time (the so-called mixing time, t_0 in Table 3) from approx. 1 ms commonly used in polymers to approximately 100 μs enhances the relative intensity of the methylene carbons in the spectrum [164], because of their shorter cross-relaxation time, e.g. in comparison with that for the aromatics. For more quantitative estimations, however, one has also to consider the proton relaxation time in the rotating frame $T_{1\varrho}^H$, as it will be done in Sect. 3.4.2.

3.4.2   Consequences for ^{13}C Cross-Polarization Measurements

3.4.2.1 Origin of the Measured Signal

From the considerations in Sect. 1.2.6.1 (cf. Eq. (50) and the conclusions immediately following) the reduction of the optimum signal enhancement due to the cross-polarization in dependence on the cross-relaxation time T_{CH} and relaxation times in the rotating frame $T_{1\varrho}^H$ and $T_{1\varrho}^C$ can be estimated. Assuming for simplicity that

$T^C_{1\varrho} \gg T_{CH}$, the optimum enhancement is obtained only in the case of $T^H_{1\varrho} \gg T_{CH}$, otherwise a drastic reduction can occur, e.g. in the case of $T^H_{1\varrho} = T_{CH}$ by a factor e^{-1} and still by .77 for $T^H_{1\varrho} = 10\, T_{CH}$ (both even for an optimal contact time).

Let us now consider concrete conditions, e.g. for polyethylene. For the polarization contact time of 1 ms it can be shown that mainly $T^H_{1\varrho}$ influences the value of the carbon magnetization M_C at the end of the contact. One can roughly estimate $M_C \approx 0.9\, M^0_H$ for the crystalline and intermediate components ($T^H_{1\varrho} \approx 10$ ms) and $M_C \approx 0.4\, M^0_H$ for the amorphous one ($T^H_{1\varrho} \approx 1$ ms).

This is connected with a stronger damping of the amorphous components by a factor of > 2. In the example considered in Chap. 4, the crystallinity was about 85%, i.e. the remaining amorphous part of the signal was less than 10%, the signal mainly measured is caused by the crystalline and the intermediate phases and a component in the $T_{1\varrho}(^{13}C)$ decay concerning the amorphous polyethylene does not occur. Moreover, in the crystalline and intermediate regions of polymers are also sites of higher mobility (defects) marked by shorter $T^H_{1\varrho}$ values and therefore more damped or even not visible in the spectrum: also structural differences can cause a change in the relaxation times and thus an influence on the signal intensity of the considered structural unit. Therefore caution is necessary in the discussion of the intensities of different structural or motional phases.

Consequently in performing cross-polarization experiments, one has to estimate the influence of the relaxation times mentioned above in order to know how strong the different parts of the sample contribute to the spectrum.

3.4.2.2 Effects on the Motional Parameters

We only consider those phases which we can measure, i.e. the crystalline and the intermediate components in the example discussed above. If the signals of the mobile defects are suppressed, then those parts of the polymer do not contribute to the spectrum, for which the motional amplitudes are relatively large. Remembering the discussions in Sect. 1.3.3 concerning the anisotropy parameter q^2, this means the absence of portions in the spectrum characterized by large values of q^2. Consequently, in such cases with different types of motions or distributions of correlation times the parameter q^2 for the remaining spectrum decreases in comparison with that determined without cross-polarization technique, e.g. using proton measurements. Furthermore, if one can assume that motions in a special correlation time region are suppressed, (e.g. in the above example the oscillations with larger amplitudes correspond to longer correlation times) then the width of the correlation time distribution decreases. An asymmetric distribution with fewer long correlation times also yields an asymmetric spectral density with a steeper decay for low frequencies (cf. e.g. Figs. 6, 7).

3.4.2.3 Effects of Spin-Flips

On the basis of Sect. 1.2.3 ([13, 14, 33], cf. [195] also) the influence of the spin-flip effects, characterized by the relaxation time T^0_{CH}, on the relaxation time in the rotating frame $T_{1\varrho}^C$ can be established. Using Eq. (45) and values for the local field in the rigid lattice from the second moment $\langle \Delta\omega^2 \rangle$ according to Eq. (6) and for the *rf*-field in the usually applied order of 30 . . . 40 kHz, one can get the results which we have summarized for practical applications: In the rigid lattice, i.e. for crystalline polymers at low temperatures, there is no influence of the proton spin–dynamics on the relaxation process for aromatic carbons or methine groups. The effect is possible for methylene carbons and dominating for methyl groups. In the presence of molecular motion as well as for amorphous regions at temperatures not too low this disturbing effect does not occur.

Because of the inaccuracy of knowledge of the local field in practice a check using the temperature dependence is useful. A growth of $T_{1\varrho}$ by decreasing the temperature is understandable only under the effect of spin–lattice processes.

3.4.3 Selective Deuteration

Although some interesting applications for selective enrichment of ^{13}C in polymers, e.g. for the carbonyl carbon in polycarbonate [188], are known, selective deuteration enables a wider field of useful applications. Extensive studies about the motional behaviour above and below the glass transition were carried out for polystyrene (PS) as well as for PS–toluene systems [267, 272, 278] (cf. Sect. 4.1.3.1). The selective deuteration of PS at the chain or at the phenyl rings, or of toluene at the methyl group or at the phenyl rings (either PS or toluene was deuterated in one of these two ways) yields detailed information concerning the motion of special groups. The shape of the solid echo spectrum for a deuterated toluene solution of PS below the glass transition temperature shows, for short pulse distances, besides the Pake line shape, a sharp line in the centre caused by the long "liquid" FID which does exist even at the time of the solid-echo maximum. The reason is that the second pulse of a solid-echo sequence is phase-shifted by $90°$ and does not influence the FID of a liquid [278].

Another example of selective deuteration is phase separation in segmented copolyesters or in polyurethanes [268]. The ^2H NMR line shape measured for segmented copolymers in which only the hard segments are labeled can be decomposed into two components. One corresponds to the signal of the homopolymer and is obviously attributed to hard segments incorporated into lamellae. The second component is very sharp and originates from a liquid-like motion of those hard segments which are influenced by the soft-segment matrix.

Selective deuteration can also support studies of molecular mobility in polymer model membranes [275] (cf. Sect. 4.1.4.1). For this purpose a methylene group of the spacer, the methyl-head group and three different methylene groups of the lipid chains have been deuterated selectively. This allows one to discuss the motional behaviour for these five different groups separately.

Investigations of order and motion in different groups of liquid crystalline polyacrylates [274, 280] (cf. also Sect. 4.1.4.1) can also be carried out for different kinds of deuteration. If the mesogenic side groups are labelled, the spectra for different angles between the director and the magnetic field (cf. Sect. 3.3.2.1) permit the determination of the order parameter S and of the width of the orientational distribution function. If a Gaussian function of the sine of the angle β between the long axis of the mesogenic group and the director is chosen for the distribution, then the distribution width parameter is a mean angle which accepts values between 10 and $20°$ for different long spacer groups [274]. Labelling the main chain or the spacer [280] it can be shown that both the spacer and even the main chain are ordered with respect to the director. The order parameters relative to the maximum value for all-*trans* chains are given quantitatively for special smectic and nematic polyacrylates where the spacer or chain, respectively, are aligned parallel with respect to the director for polyacrylates and perpendicular for polymethacrylates as the backbone of the liquid-crystalline polymer.

3.5 Highly Resolved ^{13}C-Spectra in Solid Polymers

3.5.1 General Considerations

The wide application of the high-resolution NMR in liquids demonstrates the efficiency of that method for detailed microstructural (and in recent years more and more microdynamic) investigations of polymer solutions concerning constitution, conformation and configuration as well as branching, grafting and copolymerisation etc. With the coupling of the cross-polarization and the magic-angle sample spinning techniques the possibilities are principally created to measure ^{13}C spectra in solids similar to those obtained in solutions. However in glassy polymers [14, 165] the remaining line width is roughly 10 to 100 times higher than in solutions caused by the nature of the solid polymers (provided good instruments are used). Moreover it must be stated that the experimental expenditure for the realization of highly resolved spectra in solids is much larger than in liquids. Therefore this expensive technique will only be applied if the samples are insoluble or cross-linked, if the interaction with the solvent disturbs and destroys the morphology, and if one wishes to study effects characterizing the solid state. (It is possible e.g. that the cross-polarization spectrum exhibits line splittings not observable in solutions. This is demonstrated in [200] for polyphenylenoxide, where the non-equivalence of two carbons is averaged out by thermal motion in liquids.) Because of the great similarity of the solid spectra with those in solution a lot of experience from the interpretation of the results can be used and the problems often concern the polymer chemistry. There are some extensive review articles about the high-resolution ^{13}C-NMR in solid polymers by Havens and Koenig [164], Baliman et al. [201], and Lyerla [202] which mainly reflect these chemically oriented problems. In this context the virtually infinite number of important papers by Schaefer, Stejskal and coworkers (e.g. [200], [203–205]) have also to be mentioned but as many other interesting papers cannot be cited in detail and discussed in our book. We shall consider some examples, which (i) signify the reasons for line-positions and intensities measured in solid polymer spectra and (ii) demonstrate characteristic applications.

3.5.2 Connections between the Molecular Structure and the Spectra

3.5.2.1 Conformational Influences

Differences in polymer chain conformation cause shifts of some ppm in the spectrum and can be used for the interpretation of highly resolved spectra. One characterizes the conformations of the neighbouring bonds in the backbone using the rotational isomeric state model as *trans* (anti), *gauche* (+) and *gauche* (−). The decisive influence on the chemical shift results from the so-called γ-effect, which is caused by the isomeric state of the carbon three bonds away from the carbon being observed and predicts an upfield shift of roughly 3.5 ppm for the transition of a single γ-carbon from the *trans* to the *gauche* position. This result agrees with the simplified concept that the *gauche* state causes a higher electron density at the site of the carbon under consideration, that means an increasing chemical shielding, and consequently an upfield shift of the resonance position (unfortunately described by a decrease of the chemical shifts).

This is a qualitative explanation of the signal splitting in partially crystalline polymers. In linear polyethylene [206] a downfield shift of − 2.4 ppm of the crystalline regions in comparison with the amorphous ones was observed, because of the all-*trans* conformation for the former. The opposite shift of about 1 ppm was found for polyoxymethylene [207]. In this case the crystal structure is hexagonal and exhibits a helical conformation giving rise to higher electron densities and consequently to an upfield shift of the crystalline signals. Quantitative conclusions from the polyethylene results mentioned above were drawn in papers by Möller et al. [208] and Gronski et al. [209] who used a somewhat modified γ-effect and got a value of 0.33 for the probability of *gauche* conformations in the amorphous fraction.

3.5.2.2 Other Influences

A) Conjugation

Conjugation differences also give rise to chemical shifts. In [210] for substances with comparable bonding nature and similar crystal structure it is shown that the shifts can be explained by the correlations between the degree of conjugation, the electron density and the chemical shielding as mentioned above. The results show e.g. that conjugation does occur in the polyimide backbone.

B) Configuration

Likewise configurational difference lead to chemical shifts. As demonstrated in [211] for insoluble polyacetylene, there is a 10 ppm shift downfield by changing the configuration from *cis* to *trans*. These results can, of course, again be explained in terms of changes in the electron density (cf. also [206, 213]).

C) Intensities

In connection with the measurements of isotropic spectra an additional remark has to be made. Using the cross-polarization technique, all restrictions mentioned in

Sect. 3.4 concerning the relaxation times are valid, too. However, it can be shown [214] that carbon atoms with intermolecular protons within two or three bonds will usually have signal intensities which agree with atomic ratios. The generalization of this experience to polymers is misleading because of greater distances to the next intermolecular proton and for many other reasons. The intermolecular interaction plays an important role, giving rise to a change of the cross-relaxation time T_{CH} and consequently to a distortion of the intensity ratios. That means that for comparing the intensities in highly resolved solid spectra one carefully has to examine the fulfilment of the conditions shown in Sect. 3.4.

3.5.3 Typical Applications

A) Crystal Structure

Earl and Vander Hart [166] investigating high-resolved ^{13}C spectra of several kinds of cellulose ascribed differences in resolution and intensity of the lines (here at unchanged positions) to differences in the morphology of the samples. Although the cross-polarization/magic-angle spinning technique does not reveal direct structures, suggestions can be made for the crystal structure of the samples investigated.

B) Aromatic Content

There are some examples in the literature for the determination of the ratio of the aromatic and aliphatic content in oil shale, petroleum, coal, wood etc. Resing and coworkers [206, 215] at first showed that the cross-relaxation time T_{CH} does not change by altering the aromatic content in the oil shale and that the other conditions for intensity measurements are fulfilled. The aromatic fraction of carbon atoms was then determined by measuring the two well-resolved lines according to the aromatic and aliphatic region.

C) Chemical Exchange

Lyerla et al. [216] have given an example for the proof of the chemical exchange. The principle of the method used consists in an observation of parts of the spectrum with nonequivalent carbons yielding the expected splitting for very low temperatures. However, at ambient temperature (but still in the solid state) they yield a single sharp line, which can be interpreted quite clearly as a result of exchange-averaged chemical shifts.

D) Cross-Linked Polymers

Several applications of the investigations of cross-linked polymers permit the study of the increasing rigidity of the samples following cross-linking in cured *cis*-polybutadiene [164] or of the dependence of the spectrum on the curing conditions to observe the disappearance of the acetylene end-groups in cured acetylene-terminated resins [213].

E) Conformation in Glassy and Semicrystalline Polymers

In a series of papers [209, 217, 218] Cantow and coworkers investigated micro- and macroconformations of macromolecules. The sample poly(1,2-dimethyltetra-methylene) is amorphous in the *threo*diisotactic and semicrystalline in the *erythro*-diisotactic configuration. Using methods briefly indicated in Sect. 3.5.2.1 and experiences from their high-resolution investigations in liquids, they were able to interpret the complicated spectra in terms of *gauche* $+$ and $-$ and *anti* conformations for the CH–CH, CH–CH$_2$ as well as for the CH$_2$–CH$_2$ bonds in the main chain. This was done for the glassy and for the crystalline part in the semicrystalline material and for the glassy part in amorphous samples. Here we will only mention that for the crystalline regions alternating g^+ and g^- conformations result for the CH–CH bonds; all other bonds are *anti* (which is in agreement with X-ray studies). In the glassy state populations of bonds and bond pairs and energy differences between these conformations are determined.

4 Molecular Motion in Bulk Polymers

A great number of authors use NMR to study the peculiarities of molecular motion in polymers. An impressive growth of this sphere of investigation in the past decade is due to the development of the traditional studies of the proton relaxation as well as of the investigation of other nuclei especially ^{13}C and ^2H. Of course, these investigations have different levels of development. A great number of objects in a wide region of temperatures and resonance frequencies have been studied by means of proton relaxation and the results are well described theoretically, while the investigations of other nuclei (e.g. ^{13}C) are not so frequent and the theoretical interpretation has not yet advanced very far. However, recently deuteron studies have been developed widely and well founded theoretically. Thus the investigations of the ^{13}C nuclei relaxation in solid polymers are carried out often at room temperature only, and as a rule at a single frequency. In spite of the facts just mentioned the accumulated experimental data on ^{13}C relaxation in polymers are quite sufficient to use some results of these experiments and to try to compare them with those of proton relaxation. The aim we have in this chapter is to clear up what reliable information as to the molecular motion can be taken from the multi-parameter experiment on nuclear relaxation of the different nuclei in bulk polymers. Here we also hope to reveal how this information can be obtained. For this purpose we shall use the model-free approach which is described in detail in Chap. 1. This problem will be solved by means of special examples selected according to the principle of the most complete set of experimental data on nuclear relaxation of ^1H and ^{13}C. We shall also take into account the results of other methods. Three types of relaxation transitions are known to be most characteristic for polymers. They are connected with the local and segmental motions of the backbone and with the local motions of the lateral and the end groups. Therefore we shall examine in greater detail each of these motion types on the basis of various amorphous polymers at first. Then we shall show how these motions are realized in semi-crystalline polyethylene.

We shall discuss the molecular motion within the usual classification of the relaxation transitions, denoting the glass transition by α and the transitions appearing with temperature decrease by β and γ respectively. We use these designations separately for each phase of partially crystalline polymers.

4.1 Amorphous Polymers

Before the detailed analysis of each experiment we regard the most typical temperature dependences of the nuclear relaxation times of the protons of several

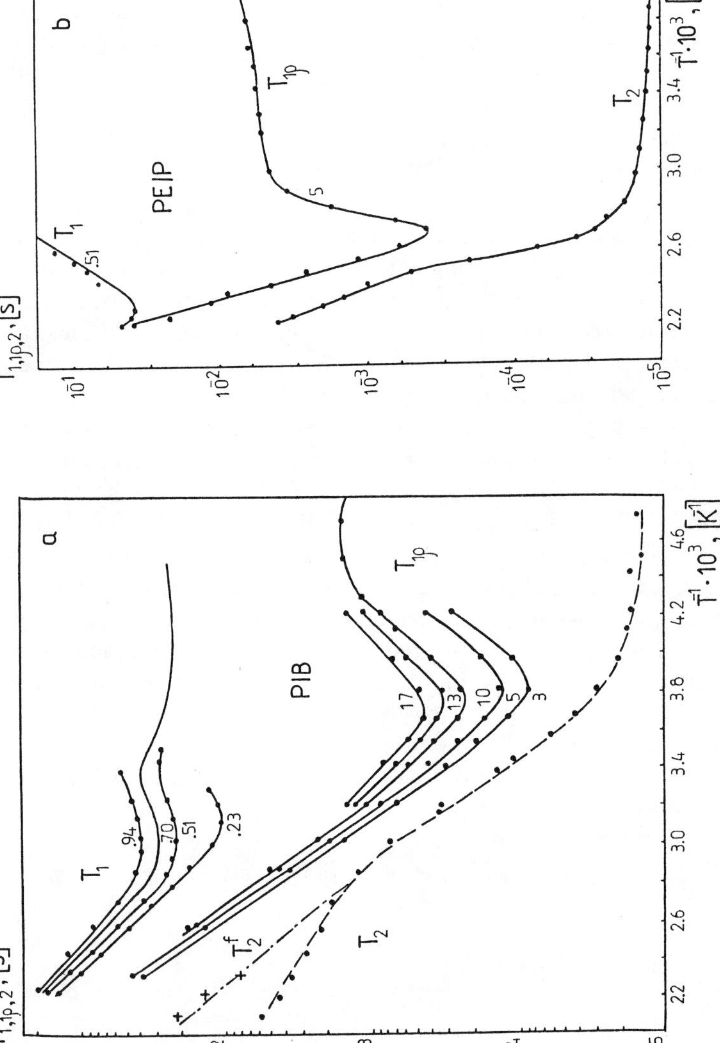

Fig. 46. Temperature dependences of the proton relaxation times T_1, $T_{1\varrho}$, and T_2 for (**a**) PIB and (**b**) PEIP. Parameters on the curves: B_1 in 0.1 mT (for $T_{1\varrho}$) and B_0 in T (for T_1)

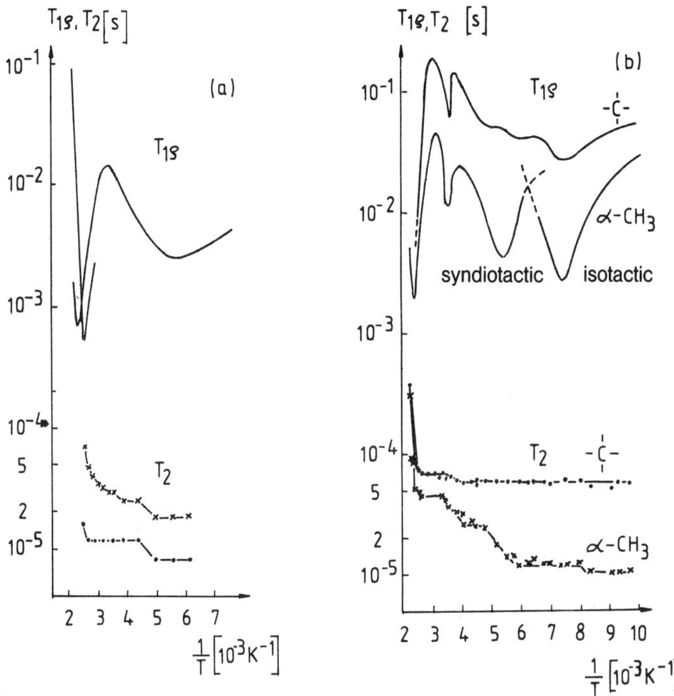

Fig. 47. Temperature dependences of the ^{1}H- (**a**) and ^{13}C- (for the α-methyl and the quaternary carbons selectively) (**b**) relaxation times $T_{1\varrho}$ and T_2 in PMMA [224, 225]. **a**: short and long component

polymers. The temperature dependences of proton relaxation times T_1, $T_{1\varrho}$, and T_2 for typical polymers without side groups (rubbers and thermoplastics) are presented in Fig. 46. Figure 47a represents analogous data for polymers having lateral groups.

The temperature dependences of the relaxation times T_1 and $T_{1\varrho}$ show two minima, one of them (at high temperature) corresponds to the glass transition (to segmental motions) while the other corresponds to local motions (at low temperatures). In the first case (Fig. 46) this is the local motion of the backbone atoms and in the second case (Fig. 47) of the lateral CH_3 groups. The comparison of the positions of the T_1- and $T_{1\varrho}$-minima for thermoplastics and rubbers shows that rubbers are the most suitable polymers to study the segmental motion (the temperature interval is 200 to 470 K) while thermoplastics are best suited for the study of local motions (the temperature interval is 200 to 350 K). Therefore we shall study the segmental motion using mainly the data obtained for rubbers, while the local motion will be investigated by means of two typical thermoplastics (PETP and PMMA).

4.1.1 Local Motion of the Backbone

We study the local motion of the backbone of polymers in amorphous PETP. Various methods (dielectric and mechanic relaxation, NMR of ^{1}H and ^{13}C) show at temperatures lower than the glass transition ($T_g = 340$ K) that in these polymers the relaxation transition, the so-called β-transition, is due to the local motion of the backbone. In this section we shall give the joint quantitative analysis of the experimental data on NMR of ^{1}H and ^{13}C. We shall be guided mainly by the data given in [36, 38, 60, 157] (^{1}H relaxation) and in [219] (^{13}C relaxation).

Figure 48 presents the proton relaxation data, which are the temperature dependences of spin–lattice relaxation times T_1, $T_{1\varrho}$, T_{1D} and T_2, and the typical decay of the transversal magnetization obtained at room temperature. Figure 49 presents the ^{13}C relaxation data, which are frequency dependences of $T_{1\varrho}$ for methylene and aromatic hydrocarbons and the longitudinal magnetization decay in the RF for CH_2 groups obtained at room temperature [219]. Next we analyse the whole experiment by means of the formalism stated in detail in the first chapter, i.e. we try to describe all the data by several generalized parameters of molecular motion connected with the shape of the correlation function, with the frequency and with the amplitude of the motion.

We mentioned in Chap. 3 that the longitudinal magnetization decays in amorphous PETP (which are due to the spin–lattice relaxation of the protons) are always exponential and are described by the relaxation times T_1, $T_{1\varrho}$, and T_{1D}. These times are much greater than the spin–spin relaxation time T_2. This allows the assumption that PETP has a uniform spin system. Hence, independent of the nature of the

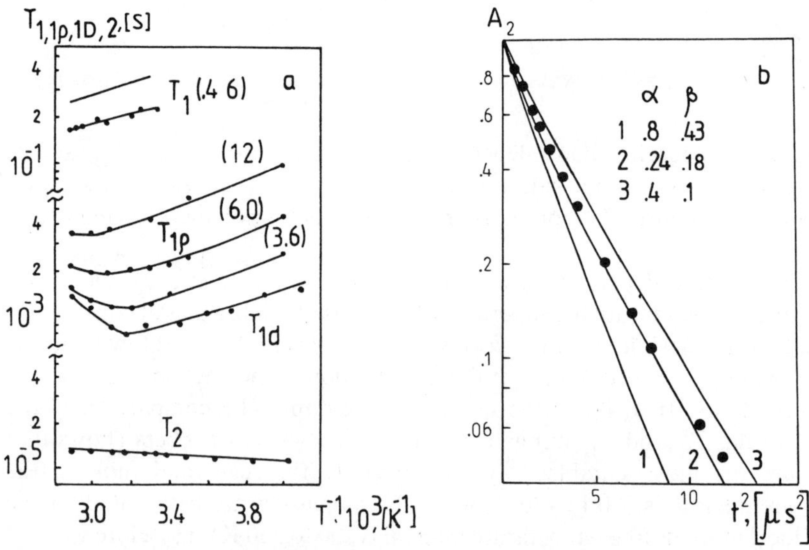

Fig. 48. Temperature dependences of the proton relaxation times T_1, $T_{1\varrho}$, T_{1D}, and T_2 (a) and FID at room temperature (b) in amorphous PETP. (b): Solid lines: calculated according to Eq. (63), circles: experiment. Parameters cf. Fig. 46

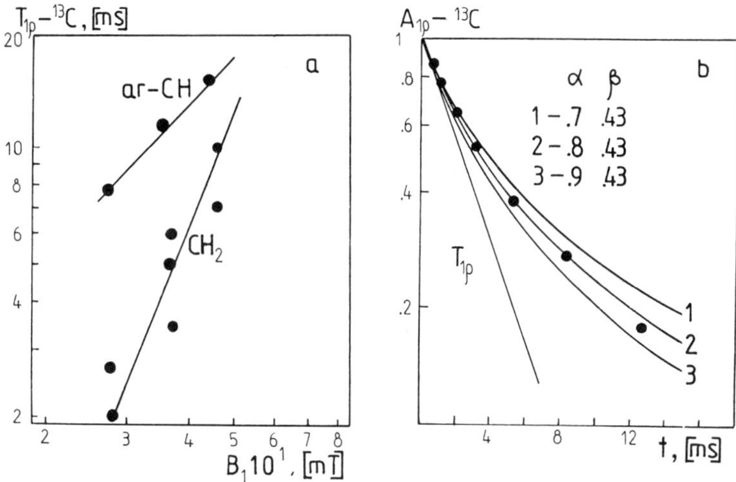

Fig. 49. Field dependence of $T_{1\varrho}-^{13}C$ for aromatic and methylene groups (**a**) and longitudinal relaxation decay in RF for methylene groups (**b**) in PETP [219]. Solid lines: calculated using Eq. (75), straight line: initial slope

spectrum of correlation times one can use Eqs. (65, 77) to analyse the spin–lattice relaxation quantitatively.

The ^{13}C relaxation involves some other aspects. It is clear from Fig. 49b that the longitudinal magnetization decays in RF are non-exponential. This undoubtedly means that the ^{13}C nuclei do not form a uniform spin system and that the signal consists of independently relaxing components. In this case the relaxation time often is calculated from the slope of the initial section of decays (cf. Fig. 49b). When the longitudinal magnetization decay is not exponential we can conclude that the system has a heterogeneous distribution of correlation times.

To get molecular parameters from the experimental data we use the expressions for the relaxation times obtained on the assumption that the distribution function of the correlation times and their spectral densities look like Eqs. (80) and (81), respectively. The analysis of the experiment was carried out in a way which describes the highest possible number of experimental data by a single set of adjustable microdynamic characteristics. Analysing the data concerning proton relaxation we mainly regard the form and the depth of the minima of the temperature dependences T_1 and $T_{1\varrho}$, the temperature dependences of the second moments, and the dependences of $T_{1\varrho}$ on the *rf* field applied.

The analysis shows that the great number of experimental data can be described by the parameters $\langle\Delta\omega^2\rangle = 8.6 \times 10^9$ s^{-2}, $q^2 = 0.5$, $\alpha = 0.3 \pm 0.05$, $\beta = 0.15 \pm 0.03$, $E_a = 55 \pm 8$ kJ/mol, while τ_0 is presented in Fig. 50 together with the data of the dielectric relaxation. It is interesting to note that the parameters α and β we have obtained are close to those won from the dielectric data ($\alpha = 0.4$, $\beta = 0.1$) [220]. Although generally the experiment is properly described by the theory (cf. Fig. 48) there are some discrepancies. At temperatures

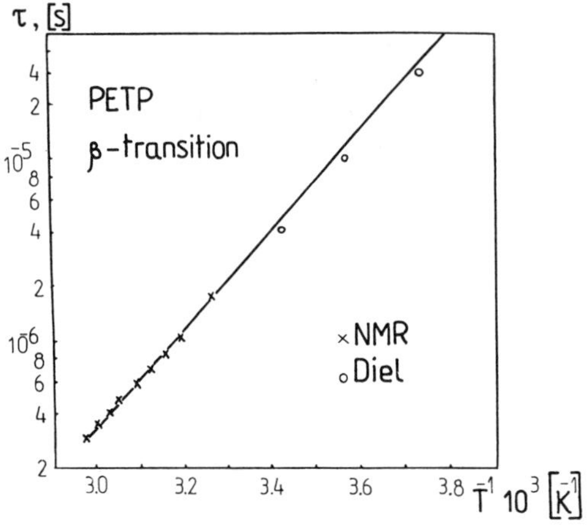

Fig. 50. Temperature dependence on the correlation time in amorphous PETP [157]

lower than the minima $T_{1\varrho} \sim B_{\text{eff}}$ holds true but not $B_{\text{eff}}^{1.15}$ which is the theoretical requirement; the quantity $\langle \Delta\omega^2 \rangle = 8.6 \times 10^9 \text{ s}^{-2}$ but not $11.5 \times 10^9 \text{ s}^{-2}$ that is observed at 4 K for the rigid lattice; the absolute values of T_1 are 1.5 times smaller than the theoretical ones. All these discrepancies can be explained easily on the assumption that an additional relaxation transition (γ) takes place at lower temperatures, which affects these parameters. The existence of such a transition was also revealed in the dielectric data [221, 222]. There are not quite enough experimental data now to make some kind of quantitative estimation. This transition can only be supposed to be characterized by a low activation energy, by a broad spectrum of correlation times and by great motion limitations (the parameter q^2 estimated from the ratio of the second moments is 0.25).

We use the parameters q^2, α, β and $\langle \Delta\omega^2 \rangle$ obtained above to calculate the transversal magnetization decay (cf. Fig. 48b) by the expressions (74) and (75) for the homogeneous and heterogeneous distributions, respectively. The figure shows that the curves calculated on the supposition of a heterogeneous distribution agree better with experimental values. This result proves the qualitative conclusion drawn from the nonexponentiality of the longitudinal magnetization decay of carbon (Fig. 49b), with respect to the heterogeneous nature of the distribution of the correlation times of the motion studied. Now we shall discuss the dependences of the relaxation times $T_{1\varrho} - {}^{13}\text{C}$ on the field B_1 that are represented in Fig. 49a. We assume that the relaxation of carbons is determined by just the same dynamic process, and we use the values represented in Fig. 50 for the analysis. At $T = 25\,°\text{C}$ is $\tau_0 = 2 \times 10^{-5}$ s that means that the condition $\omega_1 \tau_0 \gg 1$ is fulfilled for all values of the field B_1 used in the experiment. In this case according to the theory $T_{1\varrho}$ must be proportional to $B_1^{1+\beta}$ ($0 < \beta \leq 1$). Hence, the dependence $T_{1\varrho}$ of methylene groups cannot be described within the limits of our formalism ($T_{1\varrho}^{\text{CH}_2} \sim B_1^{2.5}$), while the same dependence for aromatic carbons can be described quite well since $T_{1\varrho}^{ar} \sim B_1^{1.4}$. To understand this behaviour of $T_{1\varrho}^{\text{CH}_2}$ we have to assume some additional mechanism, which has a stronger field dependence and contributes to the relaxation of the CH_2 groups. This mechanism is more effective for CH_2- than for CH-groups (the local field at the carbon sites is somewhat larger in the first case) and its effectiveness grows with decreasing B_1. Therefore one can suppose that spin-flips represent this mechanism (cf. Sects. 1.2.3 and 3.4.2.3).

Since the ratio $T_{1\varrho}^{ar}/T_{1\varrho}^{CH_2}$ decreases with growing B_1 and at $B_1 = 4.4\,mT$ reaches the theoretical value of 2, the values $T_{1\varrho}^{CH_2}$ at this field and T_1^{ar} at all fields are determined by spin–lattice relaxation processes.

We use our approach to analyse the relaxation times and magnetization decays with the expressions (76) and (74). We take into account that the dipolar interaction constant is to be diminished by $\approx 25\%$ due to the γ-transition, and the magnetization decay obtained at 3.7 mT is affected by spin–dynamics processes. The analysis yielded the following set of adjustable parameters, which is the same for both investigated groups: $q_c^2 = 0.03$, $\alpha_c = 0.7$, $\beta_c = 0.43$.

Comparing the latter with those obtained from proton relaxation data one can see that q_c^2 is smaller than q_H^2 by more than one order of magnitude and that the spectrum of correlation times is considerably narrower than the one obtained from ^{1}H. Figure 51 represents distribution functions of correlation times obtained from three experiments.

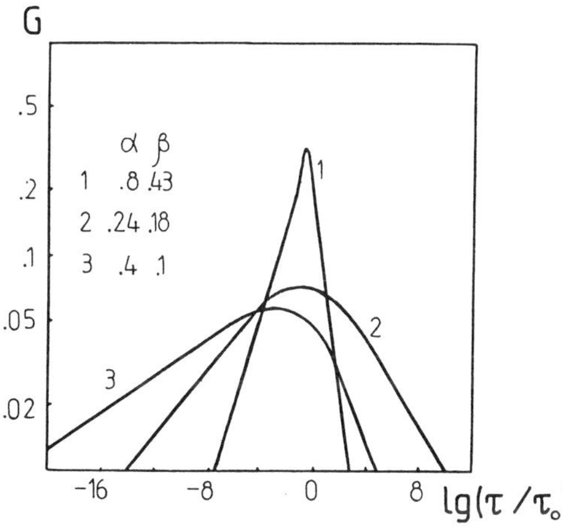

Fig. 51. Distribution functions of correlation times from ^{13}C-NMR(1), ^{1}H-NMR(2) and dielectrical data (3) for the β-transition in PETP and corresponding distribution parameters according to Eq. (80)

The smaller value of q_c^2 is most probably due to the peculiarities of the measurement of $T_{1\varrho}^c$ by the cross-polarization method (cf. Sect. 3.4.2.2). These peculiarities are connected with the loss of signal from those nuclei that have the greatest amplitude of motion. It can take place in systems with sufficiently broad distributions of motional amplitudes (parameter q^2). The same effect can also lead to a narrowing of the spectrum of the correlation times if the motion mechanism consists in certain oscillations of the nuclei since a connection between amplitude and frequency is obvious in oscillation processes (the amplitude grows with decreasing oscillation frequency).

Of course, there can be other reasons of such a narrowing of the spectrum. They can be due to the fact that in the experiments with ^{13}C we operate with a magnetically diluted spin system where the correlation function of the dipolar interaction is determined mainly by the C–H vector motion while in the case of dielectric and proton relaxation the correlation function provides additional information on intergroup dipole–dipolar interactions.

Because three different kinds of the experiment (nuclear relaxation of ^{1}H and ^{13}C, dielectric relaxation) which provide information on the motion of CH_2-, CH- and C=O-groups are described within the frame of a general relaxation process, we can conclude that all groups take part in the same type of motion. This motion is of evidently anisotropic (limited) character and is described by a broad distribution of correlation times that is of a heterogeneous nature.

Furthermore we have reasons to assume that the distribution of motional amplitudes (q^2) is connected with the distribution of correlation times.

The above results make it possible to suppose that one of the most probable mechanisms of this transition is some oscillation of the backbone nuclei. On the assumption that the motion of the internuclear vectors can be represented by the model of stochastic oscillations in a cone (cf. Sect. 1.3.4.1) and that q_H^2 yield accurate information on the motion we can easily obtain the amplitude characteristic of the motion, i.e. the half-angle of the cone from Fig. 8. In our case it is $\approx 35°$ to $40°$.

4.1.2 Motion of Side- and End Groups

The motional behaviour of side groups in polymers will be discussed here firstly for poly(methyl methacrylate) with a dominant influence of the α-methyl groups, and secondly for some aromatic polymers with a characteristic ambient influence.

4.1.2.1 Poly(methyl methacrylate)

Besides the temperature dependence of $T_{1\varrho}$ and T_2 of protons, some investigators measured those relaxation times for ^{13}C selectively using the cross-polarization method with [223] and without [224–226] MAS-technique. The latter permits the observation of the relaxation times of the α-methyl group carbons as well as of the quaternary ones, which are relatively well separated, whereas the carboxyl signal, though well separated, is too weak, and the two other signals coincide. The advantage of a measurement without MAS-technique lies in a greater variability of the temperature, which is often limited in the MAS experiments [227], (e.g. in [223] the important temperature region of the glass transition was not reached). Figure 47b shows the temperature dependences of $T_{1\varrho}$ and T_2 for protons in comparison with those for the α-methyl and the quaternary carbons. One can clearly distinguish the minima for $T_{1\varrho}$ and the corresponding steps in the T_2 curves for protons according to the dynamic glass transition at approximately 418 K and at lower temperatures in connection with the α-methyl rotation. Besides these transitions at intermediate temperatures an additional transition can be observed for carbons caused also by motional effects probably by the water adsorbed in the ester side-group. A splitting of the α-CH_3 minimum into two occurred as well. The reason for the latter effect is that the sample consists of 80% syndiotactic PMMA, the rest is isotactic. Because of the representation of only the short component in the low temperature region (whereas for higher temperatures a mean value for an e^{-1}-decay of the complicated signal is used), both minima can be observed separately. The minimum for the syndiotactic material lies at higher temperatures (188 K compared to 137 K) owing to the more restricted motion. There are some interesting conclusions from the more detailed interpretation of the experimental material. First of all it must be noted that the dependence of the other carbons on the temperature shows a qualitatively similar course, though with smaller variations of the absolute values. That means that the cooperativity of the molecular motions is very high and especially the α-methyl relaxation influences all other relaxation mechanisms very effectively. Moreover, it can be established that the spin–dynamics effects do not play any role for this interaction. From Sect. 3.4.2.3, it follows that for amorphous polymers at temperatures higher than 100 K and *rf* fields of about 4 mT the spin–flip effects can be neglected in comparison with the spin–lattice relaxation. Furthermore, the measured $T_{1\varrho}$-values exhibit an increase with decreasing temperature near 100 K at the low temperature side of the isotactic α-CH_3 minimum mentioned above. This is a proof of the domination of the spin–lattice interaction. The discussion of the quantitative curve leads to an estimation of the activation energy E_a, the width of the correlation time distribution β, and of the anisotropy parameter q^2, all for proton as well as for carbon signals. The parameter q^2 was determined from $T_{1\varrho}$ as well as T_2 measurements. The best agreement was obtained for the activation energy E_a. The results for protons and carbons, from the slope and from the frequency dependence, show an accordance with the values found in the literature of approx. 200 KJ/mol for the glass transition (which is of course an apparent activation energy only from the NMR-data, the bend of the WLF-curve becomes visible only by the addition of measurements with low frequency methods) and approx. 23 and 30 KJ/mol for the α-methyl motion in the isotactic and the stiffer syndiotactic part of the sample, respectively. The fitting of the glass transition can be done by the use of a Fuoss–Kirkwood distribution and of the WLF-equation for the temperature–frequency relation, and shows values for the distribution parameter β of about 0.5 for protons and carbons. On the contrary a somewhat broader distribution ($\beta = .25$) follows for the β-methyl motion, fitted with an Arrhenius equation, with respect to protons. However, for both ^{13}C-minima the correlation times are nearly undistributed ($\beta \approx .9$). This surprising result was verified by measurements of pure iso- or syndiotactic materials. This seems to be in contradiction to the concept of a broad distribution of local motions. We believe that a selective observation shows the methyl-group rotation with relatively weak external influences on the frequency, and that the signals are therefore determined by intramolecular interactions. The proton signal, however, represents three kinds of protons, all influenced by the α-CH_3 groups in different ways, even by intermolecular interactions, giving rise to a much broader distribution of correlation times. The anisotropy parameters q^2, derived by means of the activation energies and distribution parameters discussed above, roughly show the same part of the second moment averaged by the α-methyl-group rotation for the protons in comparison with the sum of both processes characterized by the two ^{13}C-$T_{1\varrho}$ minima. From the $T_{1\varrho}$ minimum of the glass-transition, however, for the proton measurements there follows a value of q^2 about four times greater than for carbon. The reasons for this effect lie in the use of the cross-polarization technique for the $^{13}C-T_{1\varrho}$ measurements. As described in Sect. 3.4.2.2, only parts of the sample with more restricted motions and consequently smaller values of q^2 are observed.

On the other hand, for the α-CH_3-rotations such a broad distribution of correlation times does not exist, i.e., the cross-polarization affects all carbons under consideration in the same way, and hence q^2 is nearly the same as for the protons.

Another problem of the interpretation of the anisotropy parameter q^2 is that all averaged parts of the moments following from the α-CH_3-minimum values of $T_{1\varrho}$ for protons as well as for carbons are significantly smaller than those determined by the use of the change of the second moment obtained from the temperature dependence of T_2. A possible explanation of this apparent contradiction should be a solid-like part of the signal, which occurs at the beginning of the FID and is not turned to account, and is therefore without any effect on T_2. This solid-like part can be caused by density fluctuations in the polymer with mean fluctuation times of the order of $100\ \mu s$ which leads to a strong restriction of the CH_3-rotation due to hindrances by neighboured polymer chains. Because the T_2-values are shorter than that characteristic fluctuation time, unaveraged parts of the signal were observed. On the contrary, $T_{1\varrho}$ is longer than the mean fluctuation time and therefore averaging occurs with the $T_{1\varrho}$-value of the solid-like part and at the $T_{1\varrho}$-minimum for the movable regions results in a contribution, which decreases the averaged relaxation rate. Therefore the absolute values of the $T_{1\varrho}$-minima are too high and, consequently the parameters q^2 are too small.

The NMR data discussed here can be arranged quite well in an activation diagram given by McCrum [228] in connection with dielectrical, mechanical and other NMR-data. The two points of the selective α-CH_3 minima from $^{13}C-T_{1\varrho}$ and T_2-measurements are shifted, as expected, to higher (isotactic) and lower (syndiotactic) temperatures with respect to the straight line formed by the nonselective proton values of the sample containing both iso- and syndiotactic PMMA.

4.1.2.2 Aromatic-Containing Polymers

Froix et al. ([229] and references herein, cf. [230] also) published some articles on NMR relaxation time measurements in aromatic-containing polymers (e.g. polystyrene, poly(vinyl anthracene), poly(vinyl carbazole), polycarbonate, poly(vinyl pyridine). Here we will only discuss some aspects concerning the end- and the sidegroups and the influence of air or oxygen on the NMR relaxation of these samples. It was generally found for aromatic-containing polymers that the phenyl groups form complexes with molecular oxygen giving rise to a T_1 minimum due to the dipolar interaction. Spin diffusion takes place to the paramagnetic oxygen in these atmospheres. This minimum is not intrinsic to the polymers. Without the influence of molecular oxygen, those polymers exhibit low temperature relaxations determined by spin diffusion to the end groups or to the side chains. That means that the low temperature behaviour of such samples can be extremely complicated and the intrinsic motional processes can totally be masked by the oxygen paramagnetic effect. Therefore one has to exclude oxygen, especially in aromatic polymers, in order to observe the intrinsic molecular relaxation of the polymer. There is, however, another aspect produced by these oxygen complexes with aromatic groups concerning low temperature relaxations due to the motion of the aromatic side-groups, which are not observable without interaction with oxygen because their amplitudes are too low. Interaction with paramagnetic oxygen causes an additional dipolar field at the sites of the polymer protons. Poly(vinyl anthracene) exhibits a T_1-minimum and a transition in T_2 because the oscillation of the aromatic side-chain reduces the lifetime of this interaction. The temperature dependences show activation energies corresponding to the intrinsic motions of the polymer. This is a special case where the influence of paramagnetic oxygen supports the investigations of the motional behaviour in polymers.

4.1.2.3 Generalizations

The main results of the investigation of the side-groups in polymers can be summarized as follows:

(i) There is a very large cooperativity of the different types of molecular motions in the polymers. In all temperature regions observed the selectively considered motions affect each other.

(ii) Side-groups with a great mobility are generally dominant for the relaxation process of the whole polymer and often represent relaxation sinks.

(iii) Besides a masking of the intrinsic polymer motion, oxygen complexes with side-groups can also produce signals characteristic for that intrinsic motion, which is not observable without oxygen influence.

(iv) The width of the correlation time distribution determined by selective ^{13}C-experiments for the rotation of the side-group is very small because only intramolecular interactions are effective.

(v) As a result the cross-polarization technique does not reduce (as in the case of the glass-transition) the anisotropy parameter q^2.

(vi) The anisotropy parameter, derived from $T_{1\varrho}$-measurements, seems to be smaller than that from T_2-measurements. This can be explained by density fluctuations causing restricted motions of the side-groups.

4.1.3 Segmental Motion

In this paragraph we shall analyse in detail the data concerning the nuclear relaxation in amorphous and semicrystalline polymers at temperatures higher than their melting points or glass transition temperatures. It is known that in this temperature region the relaxation times are closely connected with the micro-Brownian (liquid-like) motion, which is often called segmental. Firstly we will discuss one of the most probable versions of the phenomenological description of the temperature dependences of the correlation frequencies obtained by various methods. Then from the detailed analysis of the experiment we shall get further information concerning the parameters of the correlation function of this motion. And at last by the examination of these parameters we shall try to draw more certain conclusions as to mechanisms and models of the motion.

4.1.3.1 Investigation of the Dynamic Glass-Transition

NMR relaxation time measurements together with other (dielectrical, mechanical, thermal) investigations yield proper parameters well suited to characterize the dynamic glass-transition. One of the possibilities for a common discussion of all those different results concerning the glass-transition is the use of Donth's phenomenological theory of the glass-transition [99, 231, 233–236]. The minima of the temperature dependences of the relaxation times T_1, $T_{1\varrho} - {}^1H$, $T_{1\varrho} - {}^{13}C$, T_{2e} and of the transition of T_2 to its value in the rigid lattice yield characteristic correlation frequencies of effective motional processes $v_c = (2\pi\tau_c)^{-1}$ with the mean correlation times τ_c for a corresponding temperature. From the conditions at the minima of the relaxation times mentioned above and at the T_2-transition the values $v_0/0.62$, $2v_1$, v_1, $1/4\tau$, and $1/2\pi T_2$, respectively, follow for v_c, where $v_0 = \omega_0/2\pi$, $v_1 = \omega_1/2\pi$, and τ is the pulse distance in the MW4-cycle. The connection between the glass-transitions observed by several methods is given by the WLF-equation, which we have used in the form [237]

$$(T - T_\infty)\lg\left(\frac{\Omega}{v}\right) = (T_0 - T_\infty)\lg\left(\frac{\Omega}{v_0}\right), \tag{171}$$

where (v_0, T_0) is a reference point on the curve, and for the asymptotes of the hyperbola holds $v = \Omega$ (higher limiting frequency, if T tends to infinity) and $T = T_\infty$ (Vogel-temperature, if v tends to zero). The representation of all experimental data in a logarithm frequency-temperature plot (also including the thermal experiments, where the corresponding frequency is chosen for a cooling rate $\dot{T}$ and a mean temperature fluctuation according to [231] $v \approx \dot{T}/\delta T$) exhibits an arrangement within two WLF curves with the same asymptotes. The scaling concept contains the statements that inside those scaling limits characteristical lengths do not exist, the motional behaviour remains invariant in relation to a change of the scale, and the cooperatively rearranging regions (CRR) (the smallest subsystems which can rearrange without any influence on their environment), are the upper limitations of scaling (index a) with much greater dimensions than those at the high-frequency low-temperature limit (index b) of scaling. According to [99, 233] the corresponding volumes can be found by measuring the step height ΔC_p in the temperature dependence of the specific heat capacity at the glass-transition

$$V_a = \frac{kT^2}{\varrho(\delta T)^2 \Delta C_p}, \tag{172}$$

where ϱ is the bulk density. The temperature dependence follows from the last equation, taking into account the relation

$$\delta T \sim (T - T_\infty). \tag{173}$$

At last the volume V_b is given by

$$\frac{V_b}{V_a} = \frac{G_{1a}}{G_{1b}},\tag{174}$$

where G_{1a}, G_{1b} are the shear moduli at the rubbery plateau zone and at the glass zone, respectively.

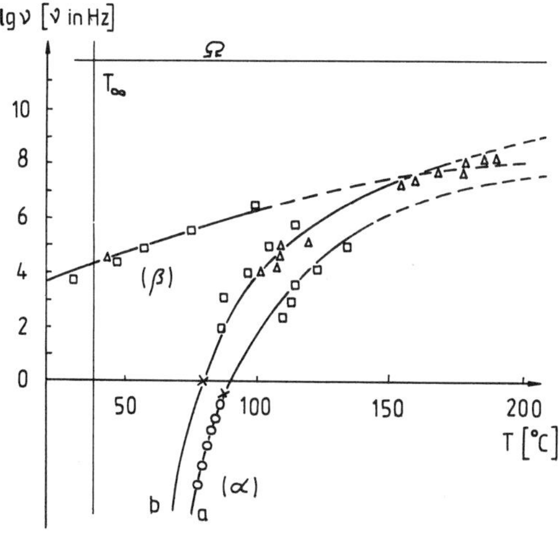

Fig. 52. Transition map for PVC with NMR ($\triangle$), dielectrical ($\square$), thermal ($\bigcirc$) and mechanical ($\times$) data [233]. Other denotations cf. text

In Fig. 52 a log v/T-diagram is presented for poly(vinyl chloride). The data position within the scaling limits confirms the more general considerations in [233] that the NMR signals are arranged at the lower limit (with respect to the volumes), the thermal and bulk properties at the upper one, and the dielectrical and mechanical measurements within the whole scaling range. The number of monomeric units which perform cooperative rearrangements and follow from Eqs. (172, 174) are about 15 for 105 to 120°C ($T_{1\varrho}$, T_{2e} data) and about 5 for 150 to 180°C (T_1 data). However, this number is not connected with special structural units, because of the phenomenological character of the fluctuation theory presented here in short. (Structural units are only important for the scaling limits). Besides the glass-transition in Fig. 52 the local-mode β-relaxation is shown, which coincides with the α-process for temperatures higher than approx. 150°C. That clearly demonstrates the suitability of $T_{1\varrho}$, or T_{2e} measurements to distinguish the α- and β-processes, whereas generally the results of T_1 do not permit such conclusions. Finally Fig. 53 shows different WLF-curves [238, 234, 239], also for PVC, but with a content of monomeric vinyl chloride of up to 15 percent by mass. The main results of the influence of VC due to an increasing molecular mobility are (i) a shift of the glass-transition to lower temperatures and (ii) a diminishing of the bend in the log v/T representation. The latter proves a decreased cooperativity of the motion because in the swollen state the volume necessary for rearrangements is reduced. Further increasing of the VC content leads to secondary relaxation

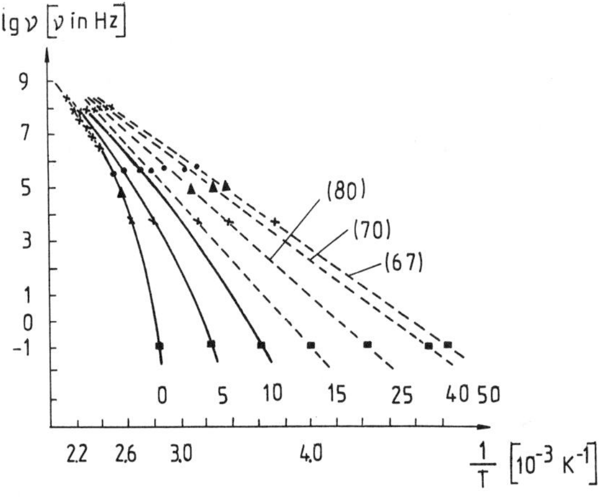

Fig. 53. Transition map for PVC with different VC ingredients [234]. Parameters on the curves: VC ingredients in mass-%, in parentheses: activation energies in kJ mol⁻¹

mechanisms due to local rearrangements, and consequently the process can be regarded as an Arrhenius process.

^{2}H NMR measurements can also contribute to the glass transition. Detailed investigations [16, 276, 278, 284] show the nature of the chain motion in amorphous atactic PS in the glass transition region. The selective deuteration technique (cf. Sect. 3.4.3) enables the characterization of the chain motion as essentially isotropic. In this amorphous polymer such highly restricted motions characterized by short flexible units like in partially crystalline PE do not exist. Additionally 180° jump motions of the phenyl rings can be observed.

Values for a relaxation map can be obtained from a T_2 representation measured by means of solid–echoes versus the reciprocal temperature. The linear behaviour above T_g changes at T_g the slope and remains approximately constant below this temperature. Analogous curves for mixed glasses PS/toluene with different compositions show details of the decrease in this glass transition region with an increasing content of toluene.

4.1.3.2 Detailed Analysis of the Segmental Motion by the Model-Free Approach

From the analysis carried out in Chap. 3, it follows that the longitudinal magnetization decays (LMD) in amorphous polymers in the majority of cases are exponential and are described by single relaxation times T_1 and $T_{1\varrho}$, while the TMD shape is more complicated and depends on the nature of the polymers and on the external conditions. We consider the spin–spin relaxation time at not very high temperatures as that time in which the TMD decreases by a factor $e(T_2^e)$. At temperatures when TMD is split into two components, we determine the two relaxation times T_2^f and T_2^s according to Eq. (135). In some cases the TMD shape will be analysed.

The data of the PE melts are exceptions. These experiments show that the longitudinal magnetization decays in RF are well described by two relaxation times $T_{1\varrho}^s$ and $T_{1\varrho}^f$ for all values of the B_1 field, while transversal magnetization decays have a super-Lorentzian shape even for the samples with $M_w/M_n \approx 1.1$. The analysis of the latter by the method described in [111, 112] (cf. Sect. 2.2.1) showed that these decays can be presented by two well separated distributions of relaxation times (T_2). We denote the relaxation times corresponding to the maxima of the distributions by T_2^s and T_2^f. The typical spectra of the relaxation times T_2 are shown in Fig. 54, while the dependences T_2^s and T_2^f taken from the maxima in Fig. 54 on M_w are presented in Fig. 55.

The typical temperature dependences of relaxation times are given in Fig. 46. In Fig. 56 the typical dependences of T_1 and $T_{1\varrho}$ on $v_1 = \gamma B_1/2\pi$ are shown for several polymers. The temperature dependences for the TMD are given in Chap. 3 (cf. Fig. 12).

The most general peculiarities of the experimental data mentioned are:

(i) There are dependences of the times T_1 and $T_{1\varrho}$ on B_0 and B_1 in the whole temperature region.
(ii) The ratio T_1/T_2 is large at the temperatures of the minima.
(iii) There are two dispersion domains at 10^3–10^4 Hz and at 10^7–10^8 Hz which are moving along the frequency scale under temperature variations.
(iv) There is an intermediate domain of 2 to 4 frequency decades of a weak dispersion between two domains of a strong dispersion.
(v) Low frequency dispersion domains and irregularities in the TMD appear at the same temperatures.

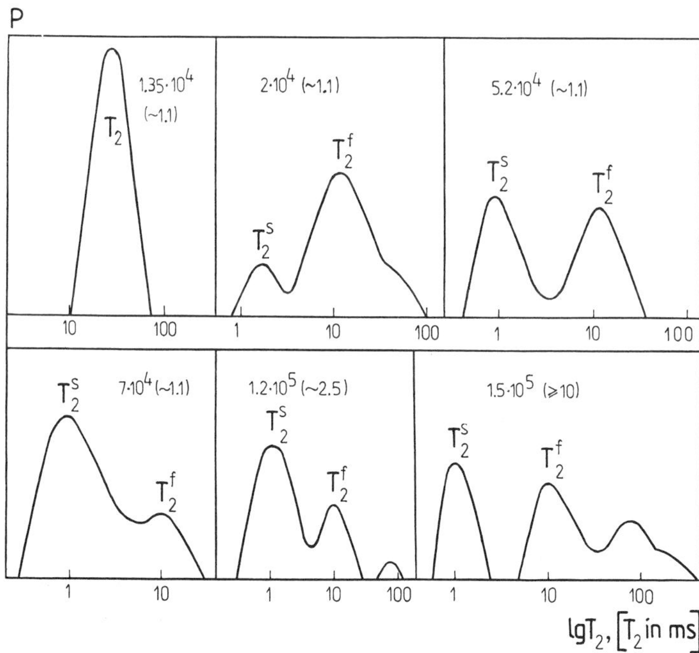

Fig. 54. Typical spectra of relaxation times T_2 in LPE melts. Parameters: molecular masses, in parentheses: ratio M_w/M_n

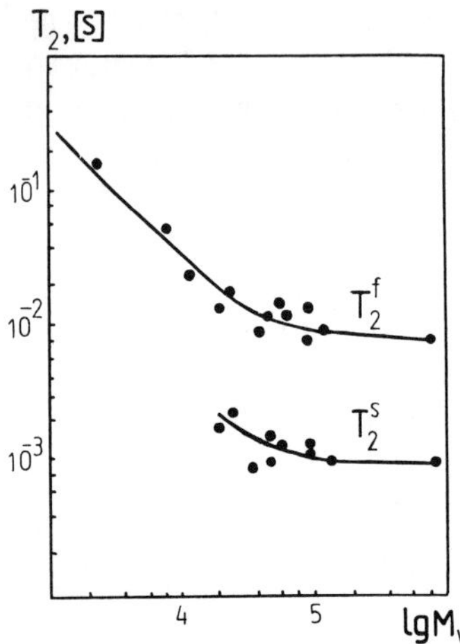

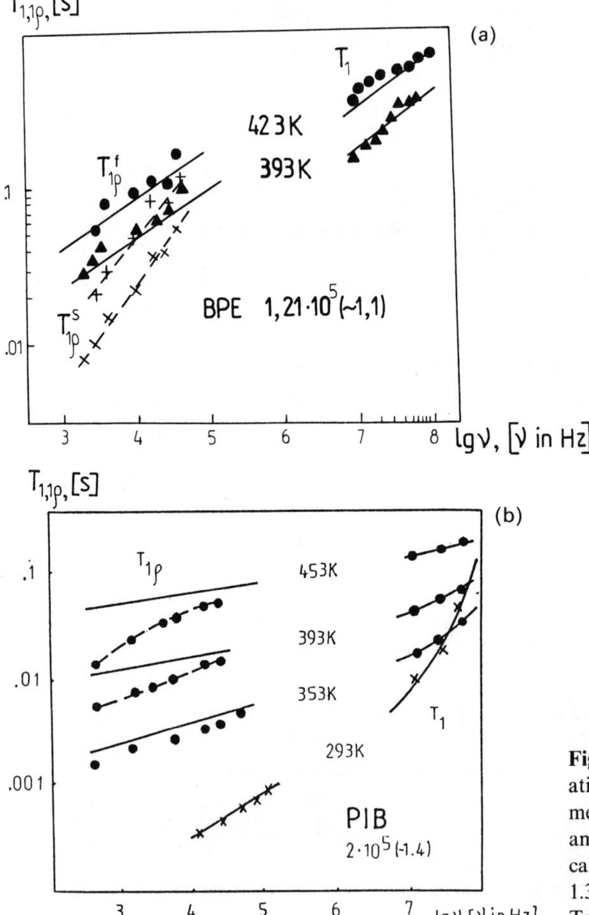

Fig. 55. Dependence of the transversal relaxation in LPE melts on the molecular masses

Fig. 56. Dispersion of the relaxation times T_1 and $T_{1\varrho}$ in (a) BPE melts [232] and (b) PIB [60]. Parameters: cf. Fig. 54. Solid lines: calculated using Eqs. from Section 1.3.3.2 using parameters from Table 7

We shall perform a quantitative analysis of our experiment by the method used in the previous paragraph. We shall assume that the motion appearing during the glass-transition is the most rapid and we designate it by the index "f". We shall designate the slower motion that is revealed at higher temperatures by the index "s" as it was done in Chapter 3. Thus, the value of the second moment measured at a lower than glass-transition temperature (T_g), is to be used for the quantity $\langle \Delta\omega_f^2 \rangle$.

Since the influence of slow motion becomes apparent only at very high temperatures, we shall firstly analyse the experiment on the assumption that only a rapid motion exists. Further the analysis will include the slow process as well if necessary.

A) Fast Motion

The detailed analysis is made using the example of PIB [58], for which the set of the experimental data on proton relaxation is the most complete. The result was that practically the whole experiment can be described by the following values of adjustable parameters: $\langle \Delta\omega_f^2 \rangle = 10^{10}\,\mathrm{s}^{-2}$; $q^2 \approx 1$; τ_{fo} is presented in the transition map (Fig. 57); the parameter of the distribution width turned out to depend on the temperature. It is presented in Fig. 58. An analogous analysis was

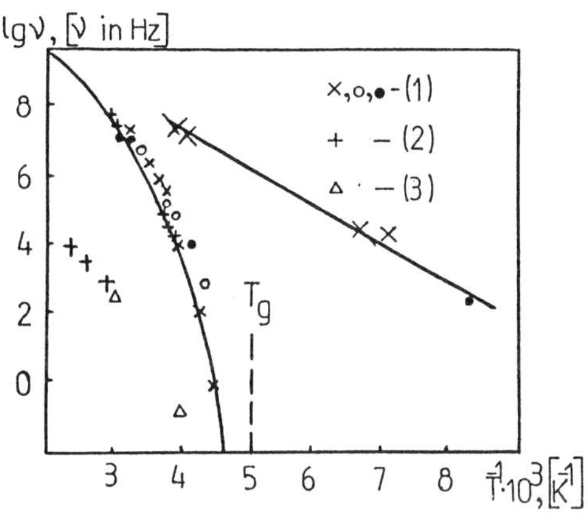

Fig. 57. Map of relaxation transitions for PIB. Data from [248] (*1*), [58] (*2*), [249] (*3*). Methods: NMR, dielectrical, mechanical (*1*), NMR (*2*), mechanical (*3*)

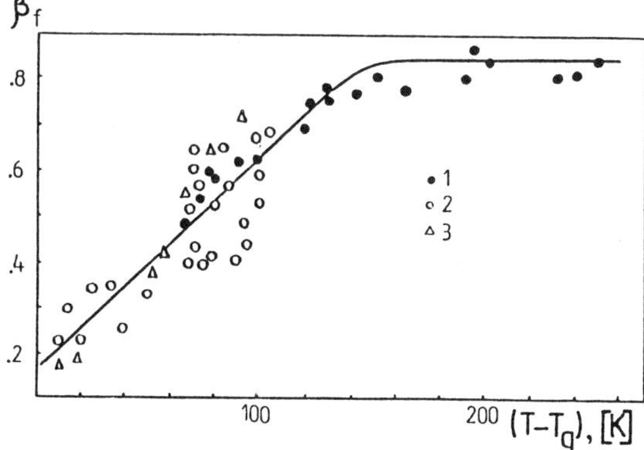

Fig. 58. Temperature dependence of the width parameter of the correlation time distribution [60] (*1*) from the dispersion of T_1, $T_{1\varrho}$, (*2*) from the minima of T_1, $T_{1\varrho}$, (*3*) data from [40]

carried out for a great number of polymers with different structures of the chain. So the data for several rubbers [60, 240–242] and thermoplastics [40, 60, 229, 243–246] were analysed. The result was that the parameter $q_f^2 \approx 1$ for all polymers, and the distribution of correlation times could be described by a symmetrical function with the parameter β_f which depends on temperature. The analysis of the data of the relaxation of the carbon nuclei made in the previous section showed that the segmental motion in the region of the $T_{1\varrho}$ minimum is also properly described by a symmetrical distribution of correlation times with $\beta_f = 0.4$ and with the anisotropy parameter $q_f^2 = 0.25$.

We can now calculate the spin–spin and spin–lattice relaxation times for PIB using the obtained values of the adjustable parameters. The calculation results are shown in Fig. 46a and in Fig. 56b. One can see from the figures that the agreement of the theoretical curves with the experiment is good except for the high temperature behaviour of T_2^e and the low frequency dispersion of $T_{1\varrho}$. Special attention should be paid to the agreement of the theoretical values T_2 and $T_{1\varrho}$ with the experimental quantities T_2^f (PIB) and $T_{1\varrho}^f$ (PE) at high temperatures.

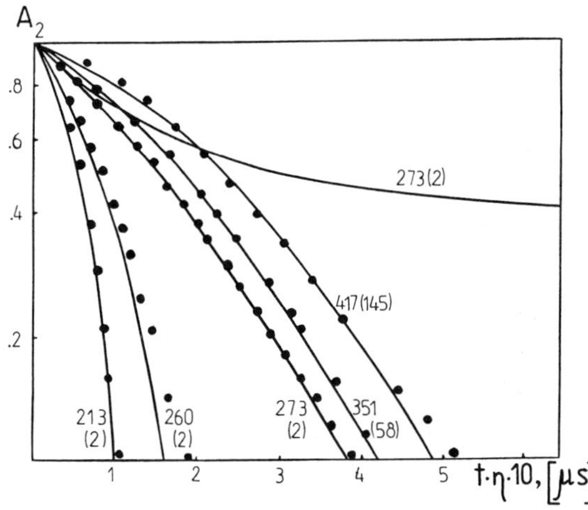

Fig. 59. Transversal magnetization decay in PIB [60]. Parameters at the curves: measuring temperatures, in parentheses: scaling factor η. Solid lines: theoretical curves for homogeneous distributions, upper solid line: for inhomogeneous distribution

Figure 59 represents the theoretical curves of the TMD for PIB calculated by Eqs. (74) and (75) for homogeneous and inhomogeneous distributions. It can clearly be seen that the experimental results are completely described by Eq. (74). This can either be due to the fact that the distribution of rapid, segmental motion is homogeneous or that there is a sufficiently rapid spin-diffusion process in the system.

Examining the temperature dependences of the parameter β_f one can see from the figure that all values β_f lie on one common curve rising monotonously as it moves away from T_g. The value β_f increases from ≈ 0.2 at $\Delta T = 30$ K to 0.8 at $\Delta T = 150$ K, at greater ΔT it remains constant.

One explanation of such a behaviour of $\beta_f (\Delta T)$ is that at low temperatures the spectrum of correlation times is caused by the segmental motion as well as by the superposition on it of the low frequency branch of the local motion. When the temperature increases the contribution of local motion decreases and the distribution τ_f only corresponds to the segmental motion. One argument for this hypothesis could be that the ratio of q_f^2 obtained from the data on relaxation of protons and carbons, was much greater than 1, which is characteristic for local motions with oscillating nature (cf. Sect. 4.1.1).

If our supposition is right the spectrum of correlation times must be of an inhomogeneous nature at low temperatures, and at increasing temperatures it becomes a spectrum caused by the nonexponential correlation function of segmental motion.

One can suppose that this question is solved quite sufficiently by the LMD shapes of ^{13}C near the glass-transition temperature. So in Fig. 60 [247] is shown that the LMD in RF for a 70% isotactic PMMA [224] at temperatures about 10 K higher than the glass transition has a super-Lorentzian shape for the α-CH$_3$ as well as for the quaternary carbons. That can be explained by an inhomogeneous correlation time distribution. Because of the great amount of scattering of the measured values at higher temperatures (the cross-polarization conditions are no longer fulfilled) the tendency to an exponential behaviour with temperature increasing could only be found qualitatively.

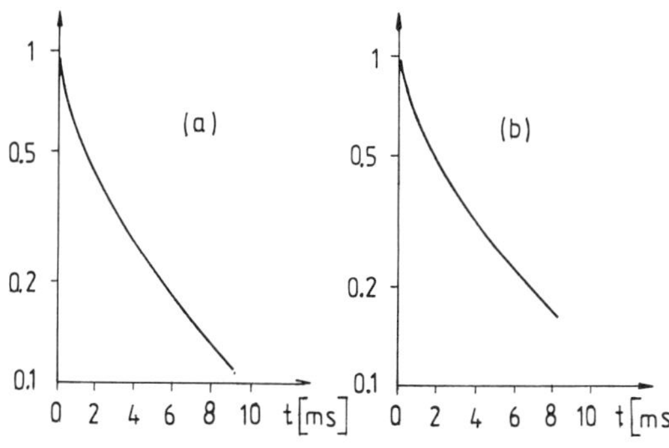

Fig. 60. Longitudinal ^{13}C magnetization decay in the rotating frame for the α-methyl (a) and the quaternary (b) carbons in isotactic PMMA at 387 K [224]

B) Slow Motion

At high temperatures the slow motion connected with the averaging of the residual dipole–dipolar interaction gains importance for the spin–spin (T_2) and spin–lattice ($T_{1\varrho}$) relaxation times. Therefore in this case one has to assume that the rapid segmental motion is anisotropic, i.e. one must keep in mind the small deviation of the parameter q^2 from 1.

We shall analyse the experiment quantitatively by the same expressions we have used for rapid motion, but we shall take into account both dynamic processes, rapid and slow. To calculate the parameters of slow motion we shall use the parameters of rapid motion obtained above.

The joint analysis of $T_{1\varrho}(\nu_1)$ dependences and TMD shape yields three adjustable parameters characterizing the slow motion τ_s, β_s, and the quantity of residual interaction $\langle \Delta\omega_s^2 \rangle = (1 - q_f^2)\langle \Delta\omega_f^2 \rangle$. These parameters are shown in

Table 7. Parameters of the molecular motion of several polymers

Polymer M_w	T [K]	β_f	τ_f [10^{-9} s]	β_s	τ_s [10^{-4} s]	$\langle \Delta\omega_s^2 \rangle$ [10^6 s^{-2}]	E_a^f [kJ / mol]	E_a^s
PIB	453	0.85	0.08	0.5	2.5	0.33		
2×10^5	393	0.85	0.3	0.5	3.5	0.8	55	17
	353	0.8	2.5	—	—	—		
	293	0.5	160	—	—	—		
BPE	423	0.64	0.25	0.5	0.015	9		
1.2×10^5	383	0.64	0.45	0.5	0.02	30	20	10.5
PBD	413	0.85	2.7	0.5	0.63	0.78		
3.2×10^5	353	0.83	9.2	0.5	1.0	1.1	26	12.5
PEO	453	0.8	1.1	0.5	0.14	0.6		
5×10^6	383	0.82	3.0	0.5	0.18	0.7	23.5	10.5
	343	0.8	7.0	0.5	0.31	0.86		

Table 8. Parameters of the slow motion in PEO for different molecular masses [60]

M_w		3×10^5			10^5			4×10^4		
T	[K]	343	383	453	343	383	453	343	383	453
τ_s	[10^{-4} s]	2.4	1.6	1.0	1.9	1.3	0.8	1.7	1.0	0.75
$\langle \Delta\omega_s^2 \rangle$	[10^5 s^{-2}]	6.2	4.9	3.5	3.4	3.0	0.27	3.0	2.3	1.0
$M_c \cdot 10^{-3}$		4.5	9.0	10.0	6.0	10.7	12.0	7.0	12.5	19.0

Table 7. Besides the table contains the correlation times τ_f and the activation energies of rapid and slow motions. We have carried out the analogous analysis for narrow-fractionated samples of *cis*-1,4-PBD and PEO [60] of different molecular masses. The adjustable parameters τ_s and $\langle \Delta\omega_s^2 \rangle$ are given in Table 8. The main results of the analysis can be formulated in the following way:

(i) Parameter $\beta_s \approx 0.5$ for all polymers.
(ii) $\langle \Delta\omega_s^2 \rangle$ is about 10^{-3} to 10^{-5} of $\langle \Delta\omega_f^2 \rangle$.
(iii) The activation energy of the slow motion is several times smaller than that of rapid motion.
(iv) τ_f is independent and τ_s is nearly independent of M_w.

C) Residual Dipole–Dipolar Interaction and Parameters of Entangled Network

We assume that the anisotropy of rapid motion is connected with the fluctuating entangled network and calculate the parameter M_c (cf. Sect. 3.1.5) using the connection of the parameter $\langle \Delta\omega_s^2 \rangle$ with the dimensions of this network according to Eq. (137). The results are presented in Fig. 61 and in Table 8.

It seems clear from Table 8 and from Fig. 61 that at approximately equal temperatures and molecular masses the values of M_c differ for all polymers and have the order of magnitude of 10^3 to 10^4, which is in quite good agreement with the data

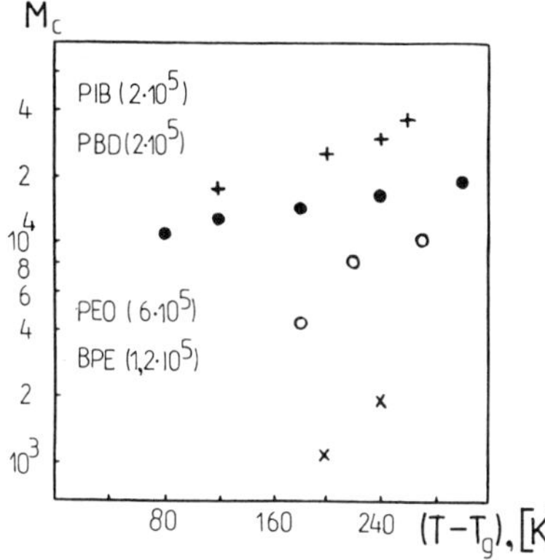

Fig. 61. Mean molecular masses between entanglements for some polymers in dependence on the temperature [60]. In parentheses: molecular masses

obtained by mechanical measurements [250]. It is interesting to note that in melts of crystalline polymers (PE and PEO) the values of M_c are lower than in the amorphous polymers (PBD and PIB), which is especially pronounced for the PE melt. The higher density of the entangled network in melts is probably due to additional supermolecular structures remaining after melting. Perhaps the memory effects observed for PE melts in [251] as well as the super-Lorentzian shape of the line (cf. Sect. 3.1.2) can also be explained by the fact just mentioned.

It is clear from the results given that M_c grows with temperature increase while it decreases with the increase of M_w and reaches a plateau for large M_w. It should be noted that the effects due to residual interaction vanish in the experiments when M_c is near $0.5\,M_w$.

The facts that the parameters of rapid and slow motions have different temperature dependences, that the correlation time of slow motion weakly depends on M_w at sufficiently great molecular masses ($M_w > 2 \cdot 10^3$), and M_c quantities and their behaviour do not contradict the concept of fluctuating entangled networks allow the conclusion that the rapid and slow motions are of different nature. Here the slow motion can be connected with the fluctuations of entangled networks.

4.1.3.3 Analysis of Motion in Polyethylene Melts
by Means of a Three-Component Model

An approach different from ours was developed in a series of papers by Kimmich [78, 252] for the analysis of the frequency dependence of spin–lattice relaxation times of protons in polyethylene melts. These authors assume that the behaviour of the proton relaxation in PE on the whole is caused by the complexity of the segmental motion. This motion was described by a three-component model based on De Gennes model [76] of reptations (cf. Sect. 1.3.4.2). For the detailed investigation of the motional behaviour in polyethylene melts in dependence on the molecular mass (and simultaneous examination of the three-component model mentioned above) measurements of the relaxation times T_1, $T_{1\varrho}$, T_{2e} and T_2 were performed [42] in narrow fractions of linear polyethylene with molecular masses from 2740 to 70100. Whereas for T_1 and $T_{1\varrho}$ spectral densities for the model derived by Kimmich [252] can be used, for T_{2e} other expressions were derived using a new formalism (and applicable for any correlation function) shown in [253, 287].

In the used temperature range from 150 to 190 °C (far off from the glass-transition temperature) it was shown by theory and experiment for a single correlation time model and for the three-component model that a nearly constant factor in the range of 1.1 ... 1.3 exists between $T_{1\varrho}$ and T_{2e}. According to Kimmich's ideas the anisotropic segmental reorientation, longitudinal chain diffusion and conformational fluctuation of the surrounding tube are characterized by the correlation times τ_s, τ_l, and τ_r, respectively, where these processes should be independent from one another and consequently, should differ in the order of magnitude of time, and depend on the molecular mass by the powers 0, 1, and 3, respectively. The experimental results showed an exponential behaviour of the longitudinal relaxation and two exponential components for the other relaxation measurements. The fitting of the experimental data is performed simultaneously for T_1 and T_{2e} or $T_{1\varrho}$, because of the small frequency ranges from 10 to 90 MHz for T_1 and from 25 to 125 kHz for T_{2e} using for the latter (i) the e^{-1}-decay, (ii) the short component, (iii) the long component, and (iv) a difference between the two.

Although the formulas for T_{2e} are very complicated expressions, numerical examinations clearly show a pure exponential decay in the ranges of correlation times considered here. Therefore the non-exponential behaviour must have other reasons. One possibility lies in the assumption that there are two kinds of molecules. The first type undergoes a hindered motion (caused e.g. by entanglements and expressed by the tube model), and yields the short component

$$\frac{1}{T_{2e}^s} = (1 - q^2)\langle \Delta\omega^2 \rangle F_1(\tau_s, \tau_l, \tau_r) + q^2 \langle \Delta\omega^2 \rangle F_2(\tau_l, \tau_r), \tag{175}$$

where the F_i are complicated functions of the correlation times. Other parts of polymer chains, e.g. at the free ends can move without hindrance and yield the long component

$$\frac{1}{T_{2e}^l} = \langle \Delta\omega^2 \rangle F_1(\tau_s, \tau_l, \tau_r). \tag{176}$$

With the parameter q^2 from the fitting of the short component a separation of $F_2(\tau_l, \tau_r)$ is also possible according to

$$\frac{1}{T_{2e}^m} = \frac{1}{T_{2e}^s} - (1 - q^2)\frac{1}{T_{2e}^l} = q^2 \langle \Delta\omega^2 \rangle F_2(\tau_l, \tau_r). \tag{177}$$

The frequency dependence of T_1, T_{2e}^l and T_{2e}^m is shown in Fig. 62 for linear polyethylene with different molecular masses.

The summary of a lot of results roughly shows the constancy of τ_s with the molecular mass and an exponent of about 3/4 with fluctuations from 0.5 to 1 instead of the desired 1 for the τ_l dependence. However, the mean value of 1.6 instead of 3 for the exponent in the τ_r dependence on the molecular mass, also with strong fluctuations, is not consistent with the theoretical predictions. Furthermore, often the fitted curves yield very similar values for τ_s and τ_l in contradiction to the

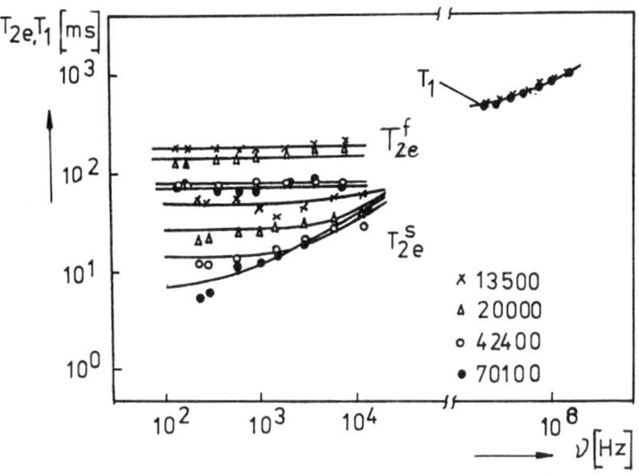

Fig. 62. Frequency dependence of T_1, T_{2e}^f, and T_{2e}^s for LPE. Drawn lines: Theoretical curve fitting. Parameters at the signs: molecular masses

theoretical assumptions. That means the three-component model for the motion in polymer melts has to be modified with regard to the component describing the longitudinal chain diffusion as well as to the dependence of the tube motion on the molecular mass. Recently Kimmich has introduced some modifications [79, 254] which are not considered in the papers discussed in this section.

Other interesting results concerning polyethylene melts investigated by several proton-relaxation techniques are described in [256].

4.1.4 Special Applications of ^{2}H NMR

4.1.4.1 Types of Molecular Motion

At the beginning we shall summarize the line shapes calculated in Sect. 1.3.5.2 for fast motion in comparison with that for a static C–^{2}H bond. We shall include a crankshaft-5-bond motion [16] representing jumps between three equidistant sites which form an angle of $2\beta_m$ with the symmetry axis. According to point (ii) in Sect. 1.3.5.2, this leads to the same result as for a rotating methyl group.

For a static C–^{2}H bond the signal has two singularities at $\pm \delta$ and two edges at $\pm 2\delta$. A rotating methyl group and a crankshaft-5-bond motion are characterized by the same line shape; however, the spectrum is three times smaller. The line shape is also retained for free phenyl-ring diffusions, but the spectrum width is even reduced in comparison with the static one by a factor 8. For a kink-3-bond motion (*trans–gauche* transitions) we observe only one singularity in the centre and two edges at $\pm \delta$. At least 180°-phenyl-ring flips lead to singularities at $\pm \delta/4$ and two pairs of edges at $\pm \delta$ and $\pm 5/4\,\delta$, respectively. However, it must be taken into account that not all C–^{2}H bonds are affected for a special type of motion. Therefore, the unaffected deuterons, as for example those in the *para* position for 180°-phenyl-ring flips, retain their static line shapes.

Of course, the real motional processes in polymers are generally more complicated and must often be expressed by suitable superpositions. For the detection of slower motions the line shapes have to be calculated according to Sects. 1.3.5.3 and 1.3.5.4. However, often the existence of only a solid or spin-alignment echo or their dependence on the pulse distances can already give important qualitative and even raw quantitative information about the motional process. As examples we shall discuss here only two types of motion occurring in polycarbonate and the mobility in liquid–crystalline polymers and in polymer model membranes.

A) Polycarbonate

Selective deuteration of the methyl and the phenyl group in polycarbonate enables one to discuss the internal motion for these two groups separately.

The chain motion can be monitored by the methyl deuterons because the averaged fgt of these lies along the C–CH$_3$ bond. These methyl signals show a Pake spectrum reduced in width by a factor of three because of the fast methyl rotation even at 380 K [16, 273]. This means the C–CH$_3$ bond is fixed, and completing spin-alignment echo measurements prove [16] that this immobility remains at least 50 ms. Detailed studies of the solid-echo spectra [16, 273] can be explained by a log-Gaussian distribution of correlation times about 2.7 decades in width and prove that the methyl rotation consists of threefold jumps about the C$_3$ axis. The line shape changes for fully and partially relaxed spectra as well as the non-exponentiality of the spin–lattice relaxation point to a heterogeneous nature of the correlation-time distribution. These facts are also valid for the relaxation of the deuterons in the phenyl rings [16]. It can be concluded, that the deuterons with different mobility are spatially separated and retain their differences at least several seconds. This heterogeneous distribution of correlation times can be explained by differences in packing in the amorphous phase leading to a distribution of activation energies. The study of such heterogeneities is in practice important because they greatly influence the mechanical properties.

The spectra for the phenyl groups can be interpreted by 180° flip motions augmented by angle fluctuations about the same axis with a rising angle for an increasing temperature. The spectrum changes from a Pake type with increasing temperatures to a type characteristic for 180°-flip motions. For an intermediate region both types can be observed simultaneously in the fully relaxed spectrum, whereas in the partially relaxed spectrum only the latter type occurs. This manifests once more the contemporaneous presence of short and long correlation times.

Another possibility for qualitative conclusions from a consideration of ^{2}H spectra is the influence of low molecular plasticizers on the spectrum of the phenyl deuterons in PC [16]. Parts of the 180°-flip type are replaced by a Pake type showing that the flip motion is blocked for a substantial fraction of monomer units by the plasticizer.

B) Polymer Liquid Crystals and Model Membranes

In addition to the discussion of the order in liquid–crystalline acrylates with isotopically labelled terminal phenylene rings of the mesogenic side groups (cf. Sect. 3.4.3) the mobility of these groups has also been considered [16, 274]. The treatment is similar to that for the phenyl groups in PC concerning the transition from the Pake type to the 180°-flip type of the spectrum. The interpretation requires log-Gaussian distributions of correlation times about 2.5 decades in width. This distribution permits not only a fitting of the line shapes depending on the temperature but also the explanation of the intensities characterized by a reduction factor mentioned in Sect. 1.3.5.3. The motional behaviour of polymer model membranes [16, 275] (selective deuteration of which was also mentioned in Sect. 3.4.3) can be interpreted using a six-site jump model for the lipid chain which describes rotations about the long axes of the molecules and conformational changes. Rising temperature leads to increasing jump rates between these six sites and to a greater number of possible conformations. For lower temperatures a two-site exchange must additionally be assumed. From a mean jump frequency introduced for the different groups, the change in mobility in comparison with the monomer membrane can be discussed. In the latter, differences in mobility for the different groups do not exist in the liquid–crystalline phase. Whereas for the lipid chains the mobility is retained after polymerization, the mean jump frequency for the head group is larger by more than two orders of magnitude. The spacer, tightly connected with the polymer main chain, has an even lower mobility.

4.1.4.2 Anisotropy of T_1

Calculations of solid-state deuterium relaxation times T_1 [283] show an anisotropy of T_1, i.e. the relaxation can be different for each orientation and consequently for each frequency. Therefore, even for an axial symmetric pattern the relaxation rates of the parallel and perpendicular edges of the Pake pattern are different [268] and the relaxation function is non-exponential. This is shown, for example, for polycrystalline alanine [268] or for poly(butylene terephthalate) (PBTP) labelled at the phenyl ring [285]. Theoretical results for the methyl relaxation can be derived [283] for jumps between three equivalent sites or for free diffusion about the C_3 axis and exhibit the following ratios for the relaxation times of the parallel and the perpendicular edges of the powder pattern: for threefold potential 2:1 in the extreme narrowing case and 4:3 in the limit of slow motion and for uniform rotation 1:1 and 3:2 in these two cases. The use of these ratios for extreme narrowing, whether there are differences in the relaxation time for the two edges or not, to distinguish between these two types of methyl group motion, is rendered more difficult because the motion often lies between these two limits and additional backbone motions can also influence the relaxation process. In all other cases, i.e. if the angle θ (cf. Eq. 38) is not equal to 0 or 90, relaxation also depends on ϕ (even for the axially symmetric case). This also holds true for jumps between two non-equivalent sites for all values θ except $0°$. The estimation of the correlation time can only be carried out here for $\theta = 0°$, whereas it is also possible for methyl groups for $90°$. However, investigations in PBTP [285] yield a smaller value than that obtained by line shape analysis for pure $180°$ flips. This indicates that rapid librational motions additionally influence the relaxation process.

The effect of tacticity on T_1 was investigated for PMMA labelled at the α-methyl group [268]. Whereas at 183 K, where the backbone motions are frozen in, differences in T_1 for syndiotactic and isotactic PMMA do not occur, they are noticeable at room temperature and still larger at a temperature between the glass transitions of the two samples. The smaller T_1 for the syndiotactic sample exhibits the greater hindrances for the α-CH$_3$ rotation. The T_1 for the ester methyl group is, of course, larger by two orders of magnitude than those for the α-methyl group, thus reflecting the far faster rotation rate of the side group.

4.1.4.3 Concluding Remarks to ^{2}H NMR

For the many possibilities of ^{2}H NMR for the investigation of structure and mobility in polymers we have tried here only to give some basic ideas about the methods (Chapters 1, 2) and some examples of characteristic applications (Chapters 3, 4). For more details we refer the reader to some reviews and other important articles by the leading researchers of these developments, Spiess and Sillescu [16, 17, 171, 172, 196–199, 276, 284].

It seems to be useful to summarize some special aspects originating from the features of the quadrupolar interaction with the fgt and from the recent experimental and methodological developments which make this technique promising. Profitable features of deuteron resonance are [16]:

  (i)   the entirely intramolecular character of the interaction mentioned above,
 (ii)   the sensitivity of the line shape for the discrimination of different types of motion,
(iii)   the wide timescale of the investigated motions in the orders 1 Hz up to 10^{10} Hz realized by different experimental techniques (line shape investigations, solid–echo, spin–alignment echo and spin–lattice relaxation time measurements),
 (iv)   the good separation of signal components characterizing fractions of different mobilities because of the small spin-diffusion influence and
  (v)   the high selectivity of this method, especially by the use of selective deuteration.

On the other hand, there are some drawbacks:

  (i)   the need for isotopically enriched samples,
 (ii)   the insensitivity to translational motions and
(iii)   we consider as a disadvantage the necessity for indirect interpretation in most cases, i.e. for theoretical calculations of different models and comparison with the experiments.

However, deuterated samples often are available in connection with neutron-scattering methods, and if the basic software is once kept ready the modifications can be done relatively fast. Therefore, one can establish in summary that deuteron NMR represents a very powerful method, not yet finished in its development, for the investigation of order and mobility in polymers.

4.2 Molecular Motion and Relaxation Transitions in Solid Polyethylene

The molecular motion in polyethylene has been investigated in a great number of studies using different methods. The most widely used methods were mechanical, dielectrical and NMR relaxation. However, in spite of the intensive study, the problem of the relaxation transitions in partially crystalline PE has not been solved completely and is still a point at issue. This is because there is neither complete agreement concerning the experimental data obtained by various methods and authors, nor clarity in their interpretation. In our view, one of the main reasons for this situation is an ambiguity of the attribution of the transitions to structural phases of the polymer.

The NMR-method can now be supposed to be probably the only method which can completely solve the problem of identification of the relaxation transitions in partially crystalline polymers. That is because the NMR-signal and its parameters are sensitive, firstly, to the heterogeneity of the sample and, secondly, to the frequency and amplitude characteristics of molecular motion. It is clear at the same time that it is very difficult to use all potentialities of this method, and it is not always possible at present. In our laboratories the systematic investigation of the magnetic relaxation of ^{1}H and ^{13}C nuclei was carried out for solid linear PE to demonstrate all the possibilities which this method offers to obtain information on structure and molecular motion in PE [291, 292]. These investigations show that it is necessary and sufficient to use the three-phase model of the polymer structure to describe the shapes of decays of transversal and longitudinal magnetizations.

This section generalizes the experimental results and considers them from the point of view of the simplest mechanisms of molecular–kinetic models of motions. We shall mainly concentrate on the joint quantitative examination of the results of ^{1}H and ^{13}C relaxations.

The majority of the results that are discussed here, was obtained with nearly identical linear PE samples, which had a very similar thermal prehistory. (They are crystallized from the melts and annealed at 120 °C). The results concerning the shape of the free induction signal and transversal relaxation times were obtained mainly in [134, 135], concerning spin–lattice relaxation times within the laboratory frame (T_1) in [116, 150, 194, 255], within the rotating frame $(T_{1\varrho})$ in [115, 257], within the dipolar frame (T_{1D}) in [153], and were discussed in detail in Chap. 3. All the experimental data on temperature dependences of nuclear magnetic relaxation times $(T_1, T_{1\varrho}, T_{1D}$ and $T_2)$ obtained in these studies are given in Fig. 63 and in Table 9.

4.2.1 Qualitative Analysis of the Experiments and Attribution of the Relaxation Transitions

Let us assume that each phase has k transitions. If we designate the main transition at the highest temperature (glass-transition or melting) in each phase by α_i where i

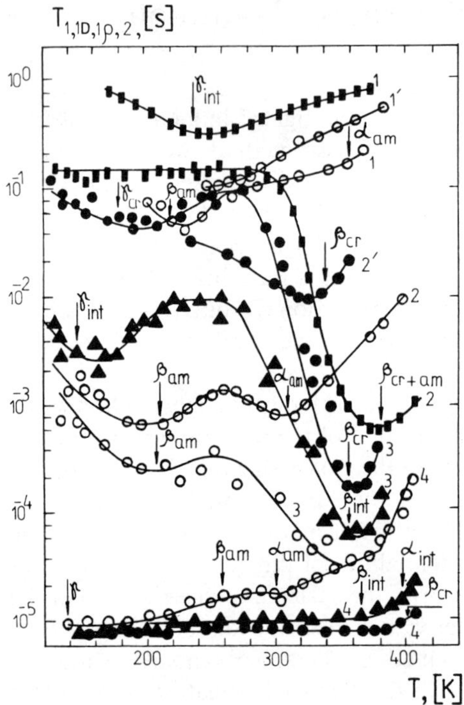

Fig. 63. Temperature dependence of the ^{1}H-
and ^{13}C-relaxation times $T_1(1)$, $T_{1\varrho}(2)$, $T_{1D}(3)$,
and $T_2(4)$ for LPE with a molecular mass
3×10^5. Prime: ^{13}C [255, 257, 292],
without prime: ^{1}H [60, 291]

Table 9. Parameters of the longitudinal magnetization decay for ^{13}C (T_{1i}, P_{1i}) in
LPE (Hostalen, annealed 4 h at 128 °C) [194] and for ^{1}H (P_{2i}) in an analogous
LPE sample [134]. T_1 (theoret.) cf. text. (§—averaged value of cr 2 and ncr 2)

| Phase (^{13}C) | P_{1i} | T_1 [s] | | P_{2i} | Phase (^{1}H) |
		exper.	theoret.		
cr 1	0.66	474	67	0.7	cr
cr 2	0.06	22			
		3.6§	3.2	0.12	int
ncr 2	0.06	2			
ncr 1	0.22	0.165	0.14	0.18	am

is the phase index, transitions that become apparent with temperature decrease by
β_i and γ_i, then k means α, β, γ, and i: cr, int, am.

4.2.1.1 Relaxation of Protons

A) Amorphous Phase
This phase presents two minima for each temperature dependence of spin–lattice
relaxation, and at corresponding temperatures the variations of the behaviour of

the times of spin–spin relaxation can also be observed. In this case at temperatures lower than 300 K the FID shape is always Gaussian, while at temperatures higher than 300 K it becomes exponential. Hence, the high temperature minima in spin–lattice relaxation correspond to the transition determined by liquid-like motion (glass-transition) which we designate by α_{am}. Low temperature minima correspond to the transitions determined by the local motions of the small sections of chains (β_{am}). Minima of the spin–lattice relaxation lower than 160 K were not observed in our experiments. At the same time in the study of branched PE at 140 K a minimum was revealed in T_1 that could certainly be attributed to the molecular motions in the amorphous regions. The existence of the low temperature transition in the amorphous phase is confirmed by the fact that the second moment $\langle \Delta \omega^2 \rangle_{am}$ at 150 K is by 1.5 to $2 \times 10^9 \, \mathrm{s}^{-2}$ smaller than the theoretical moment calculated for the rigid lattice. These data suggest that the γ_{am}-relaxation transition which is connected with the local motions in the amorphous phase too, also appears in the magnetic relaxation.

B) Crystalline Phase

The temperature dependences T_{1Dcr} show two minima, one at 200 K and another at about 360 K, while the temperature dependences of spin–spin relaxation times yield a growth of T_{2cr} above 360 K. The minima are observed at temperatures lower than the melting temperature and the growth of T_{2cr} proceeds without the signal shape becoming exponential. So the transitions designated by γ_{cr} and β_{cr} that correspond to these minima are determined by local motions.

C) Intermediate Phase

The temperature dependences of T_{2int} have two peculiarities; gentle growth of T_{2int} at 340 to 400 K proceeding without signal shape transformation and steep growth of T_{2int} at temperatures higher than 400 K accomplished by signal shape transformation from Gaussian to exponential. Hence, one can conclude that the main transition at 400 K is connected with the appearance of the liquid-like motions (α_{int}). Thus the high temperature minimum in T_{1Dint} observed at 360 K and corresponding to gentle growth of T_{2int} can be attributed to the β_{int}-transition, while the low temperature minimum in T_{1Dint} at ≈ 150 K corresponds to the γ_{int}-transition.Comparing the position and the depth of the minimum in $T_{1\varrho(cr+int)}$ with the minima in T_{1Dcr} and T_{1Dint} one can be certain that this minimum is mainly determined by the relaxation in the intermediate phase and relates to the β_{int}-transition.

The minimum in $T_{1\varrho(cr+int)}$ reflects the position of two minima that are close together with respect to frequency and temperature of the minima and corresponds to the transitions β_{cr} and β_{int}. The results of the attribution of the transitions to structure phases made above together with the data on mechanical and dielectrical relaxations are given in Fig. 64 as a transition map. It seems clear from Fig. 64 that in the experiments with the nuclear relaxation of protons in PE at temperatures lower than T_m, eight relaxation transitions are revealed and identified, three of which are in each noncrystalline- and two in the crystalline-phases.

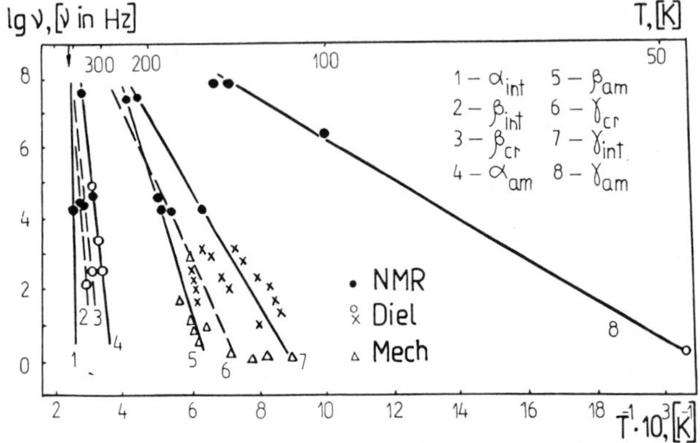

Fig. 64. Map of relaxation transitions for LPE. NMR data from Fig. 63, other methods cf. [259, 260]

4.2.1.2 Relaxation of Carbons and Deuterons

One can see from Fig. 63 that the minimum appears at a temperature of 333 K in carbon $T_{1\varrho}$ data while in T_1 data it appears at 246 and 263 K for carbons and deuterons [258], respectively. The analysis of the cross-polarization mechanisms made in Chap. 3 for PE showed that the quantities $T_{1\varrho}$ given in Fig. 63 relate to the crystalline phase only. On the other hand, from the positions of the minima $T_1-^{13}C$ and T_1-^2H, one can conclude that they are determined mainly by the relaxation of nuclei of the amorphous phase. Correlation times obtained under the conditions of minima ($\omega\tau_0 \approx 1$) lay well on the corresponding curves in the transition maps. It should be noted that this attribution of the T_1 times of the ^{13}C and ^2H nuclei does not agree with that of the authors of these experimental results.

In Table 9, the data of the phase structure of the LPE sample that is considered in this chapter are given together with the data of $T_1-^{13}C$. It can be seen that the three phases of this sample correspond to the four components of the spin–lattice relaxation of ^{13}C, if we assume that the intermediate phase from the ^1H data corresponds to the sum of the components cr2 and ncr2 from ^{13}C. Hence, the phase structures of these samples are comparable and the latter are most suitable for the joint analysis of the molecular motion. Examining the transition maps one can clearly see that at room temperature at the frequencies comparable with the resonance frequency of ^{13}C nuclei (15.4 MHz) there are motions in the crystalline and intermediate phases corresponding to γ_{cr}- and γ_{int}-transitions and in the amorphous phase corresponding to the β_{am}-transition. It should be noted that the longitudinal magnetization decays in RF observed in [257] were very close to exponential decays.

4.2.2 Determination of the Microdynamic Parameters

We shall analyse the experiment quantitatively as before by the model-free approach using it for the analysis of the transitions separately in each phase. Since we cannot determine the symmetry of the distribution accurately enough, we assume that all transitions are determined by symmetrical distribution functions ($\alpha = \beta$ in Eqs. (80, 81)), i.e. by a Fuoss–Kirkwood function. As a result of the analysis of the proton relaxation data we have obtained four adjustable parameters for each transition (q_{ik}^2, β_{ik}, E_{aik}, τ_{oik}). Table 10 and Fig. 64 represent the parameters. However, it should be noted that the accuracy of the determination of the parameters for each transition in partially crystalline polymers is many times less than that for the analogous transitions in purely amorphous polymers, and the values given in Table 10 are of approximate character.

Table 10. Characteristics of relaxation transitions in LPE [60]

Transition index	α	α	β	β	β	γ	γ	γ
Phase index	int	am	cr	int	am	cr	int	am
E_a^{ik} [kJ mol^{-1}]	209	209	125	125	59	29	29	12
β_{ik}		0.15	0.45	0.47	0.15	.07	.3	.33
q_{ik}^2	1	0.9	0.15	0.56	0.5	.003	.03	.04

Let us note the main peculiarities of the results obtained:

(i) In the region of high- (≈ 370 K) and low- (≈ 180 K) temperature transitions, observed in linear PE by mechanical methods (cf. e.g. [259, 260]), transitions of different nature taking place in various phases are revealed.

(ii) The parameter q_{ik}^2 for all α-transitions is close to 1, while for β- and γ-transitions it is smaller than 1. This parameter decreases in the sequence of $\alpha \rightarrow \beta \rightarrow \gamma$-transitions, while for β- and γ-transitions it grows at the transition from the crystalline to the amorphous phase.

(iii) The parameter of the spectrum width β is very small (<0.2) for α_{am}-, β_{am}- and $\gamma_{cr, int}$-transitions and tends to increase after the transition to less ordered systems and with growing temperature. For $\beta_{cr, int}$-transitions it is close to 0.5.

(iv) In the sequence of transitions $\gamma \rightarrow \beta \rightarrow \alpha$ there is a growth of the activation energy from ≈ 12 to 210 kJ/mol.

Now we shall carry out the analysis of the experiments with ^{13}C and ^{2}D taking the results of ^{1}H relaxation as a base. Firstly we shall calculate the theoretical values of the relaxation times T_1 for ^{13}C and ^{2}H using the parameters of the corresponding transitions presented in Table 10 and in Fig. 64. We shall perform the calculations on the assumption that $T_{1\,min}$(^{2}H and ^{13}C) is determined by the β_{am}-transition with the parameters ($\beta_{am} = 0.15$–0.3, $q^2 \approx 0.6$–0.7), while T_{1cr}, T_{1am} and T_{1int} (at room temperature) are determined by γ_{cr}, β_{am}, γ_{int}-transitions with the following sets of parameters:

$$\gamma_{cr} \quad (\beta = 0.07, \quad q^2 = 0.003, \quad \tau = (0.5\text{–}5) \times 10^{-9}\,\text{s}),$$

$$\beta_{am} \quad (\beta = 0.2 \,, \quad q^2 = 0.7 \,, \quad \tau = (4\text{–}8) \times 10^{-11}\,\text{s}),$$

$$\gamma_{int} \quad (\beta = 0.2 \,, \quad q^2 = 0.03 \,, \quad \tau = (1\text{–}2) \times 10^{-10}\,\text{s}).$$

The results are summarized in Fig. 63 and in Table 9.

It can clearly be seen that all the examined data of ^{13}C, ^{2}H and ^{1}H can be coordinated with each other and described quantitatively by the same sets of adjustable parameters.

It is interesting to note that Schröter [194] for $T_{1\,cr}$ and $T_{1\,int}$ measurements used the cross-polarization method but a decrease of q_c^2 was not observed. It is evidently due to the fact that spin–lattice relaxation times of protons $T_{1\,Dcr}$ and $T_{1\,Dint}$ at room temperature are large in comparison with the time of contact between spin systems of ^{13}C and ^{1}H nuclei (cf. Fig. 63).

Let us consider the experiment with $T_{1\varrho}$ of carbon. According to our attribution $T_{1\varrho}$ in the temperature region of the minimum is determined by the β_{cr}-transition. Let us do the same calculation of the $T_{1\varrho}$ values in dependence on the temperature using the parameters $q^2 = 0.15$, $\beta = 0.5$, $E_a \approx 120\ \mathrm{kJ/mol}$, and τ from the transition maps. To coordinate the two experiments in such a way appeared to be impossible. Another set of parameters is necessary to describe the experiment with $T_{1\varrho}{-}^{13}$C. The best description can be obtained at $\beta = 0.9$, $q^2 = 0.006$.

The narrower spectrum and the smaller parameter q^2 for ^{13}C as compared to ^{1}H measurements is quite analogous to that observed in amorphous PETP during the study of the local motion of the backbone (cf. Section 4.1.1). Hence the β_{cr}-transition in PE is characterized by a sufficiently broad distribution of q^2 and the amplitudes of the motion are closely connected with their frequencies. This shows that the oscillation process is the mechanism of the motion. But this conclusion contradicts the sufficiently narrow distribution of correlation times observed in proton relaxation and the exponentiality of the magnetization decay of carbon nuclei in RF.

In the next paragraph we shall try to draw a self-consistent picture of molecular motion in PE from the analysis of all the data obtained in this chapter.

4.2.3 Discussion from the Viewpoint of the Models of Molecular Motion

To simplify the problem we shall assume that all our results are obtained for three hypothetical samples of PE that are an ideal crystal (crystalline phase), a defective crystal (intermediate phase) and a purely amorphous sample (amorphous phase). Such an examination is possible as long as the questions of the influence of the phase structure on the parameters of the molecular motion are not concerned, and provides the chance of operating with the data obtained from purely amorphous polymers.

As it has been discussed above, in each phase of PE there are two types of transitions connected with the anisotropic local motion, and the intermediate and amorphous phases have one additional transition, each connected with the isotropic liquidlike motion. Each of these transitions results in a minimum in the temperature dependences of spin–lattice relaxation times, and can be characterized by the parameter of anisotropy and by distributions of correlation times. The existence of minima in the temperature dependences of the spin–lattice relaxation clearly shows that the molecular motions determining these transitions are stochastic, i.e. their correlation function must be represented by a sum of exponential functions.

Principally there are a great number of various motion models leading to such correlation functions. These can roughly be divided into two classes based on different concepts. One of them, the concept of the accumulation of rotation-oscillation motions of the atoms of the chain is based on the existence of stochastic-torsional oscillations appearing because of the anharmonicity of the normal torsional oscillations (cf. e.g. [261, 266]), and the other is the concept of the diffusion of defects. In the first the correlation function is represented by a sum of exponential correlation functions of each relaxator; the width of the spectrum is determined by the acoustic branch of the normal oscillations, which as the calculation shows is very broad (cf. e.g. [262]). For the model of defects the solution of the diffusion equation leads to a nonexponential correlation function (cf. Eq. (92)).

We shall analyze the results further taking into account both of these possibilities. Since we consider the model of torsional oscillations to be more basic than the model of defects we shall use the latter only if the mechanism of oscillations for some reasons is not enough.

There are two local transitions (γ and β), which occur in different temperature regions, have a different ability to average the dipole–dipolar interactions and are characterized by different activation energies. The existence of these transitions within the concept of the torsional oscillation accumulation can be presented in the following way:

The γ-transition corresponds to the appearance of the relaxation modes which are connected with the excitement of the oscillation motion of the individual noninteracting CH_2-groups. The oscillations of these groups have sufficiently small mean amplitudes and require a small free volume only. Such oscillations of CH_2-groups may have an amplitude of $20°$ and activation energies connected with the formation of the local free volume that are estimated in [261] as 40 to 50 kJ/mol. The efficiency of such motions in NMR is very small and obviously depends on the amplitude of the oscillations.

The β-transition corresponds to the relaxation modes connected with the simultaneous excitement of several neighbouring methylene groups. Naturally, such excitements will take place very seldom, require a greater free volume and, hence, a greater activation energy, but they can in turn have a greater angle due to a cooperativity of the motion. If an angle δ of such oscillations comes close to $90°$ they cannot be distinguished from complete rotations since the efficiency of these two motions is equal.

Small quantities of q_γ^2, their growth with the decrease of the order of the system ($q_{cr}^2 < q_{int}^2$), the small value of β_γ increasing for the transition from the crystalline to the amorphous phase, and sufficiently small $E_a \approx 30$ kJ/mol indicate that γ_{cr}- and γ_{int}-transitions are determined by an oscillation mechanism. In this case using the connection between the parameter q^2 and the angle δ for the oscillation model in a cone (Eq. (90)) we get for γ_{cr} and γ_{int} the angles $\approx 3°$ and $15°$, respectively.

The characteristics obtained for β_{am}-transitions ($q^2 \approx 0.5$ to 0.7), the broad spectrum of correlation times ($\beta \approx 0.15$ to 0.3), nonexponentiality of decays of the longitudinal carbon magnetization in the RF for PETP, and narrowing of the spectrum and decrease of the q^2 quantity observed on cross-polarization conditions undoubtedly show that this transition is also due to the rotation–oscillation motions. Here the amplitude characteristic of such motions is the angle $\delta \approx 40$ to $50°$ and $E_a \approx 60$ kJ/mol. This completely corresponds to the concept of the torsional oscillations.

In the case of β_{cr}- and β_{int}-transitions the sharp narrowing of the spectrum ($\beta \approx 0.5$) and the exponentiality of the magnetization decays in RF of carbons is at variance with such a model of motion. On the other hand, the narrowing of the spectra and the decrease of q^2 in the experiments with cross–polarization support the concept of torsional oscillations.

It is possible to harmonize these seeming contradictions in the following way: Let us assume that in crystals at high temperatures cooperative oscillations appear analogous to the oscillations in the amorphous phase (β_{am}-transition), but due to the high regularity of the structure they take place much more seldom, at higher temperatures, and with a higher activation energy E_a. For the time of the experiment (the spin–lattice relaxation time) not all the groups of the chain will be excited and produce an effective relaxation mechanism. Such excited states can play the role of defects which can be characterized by the concentration and different amplitudes (distribution of q^2).

The nuclear magnetic relaxation will be realized through these defects. In the case of proton relaxation the magnetization transfer can be realized by the spin–diffusion process as well as by the motion of the defects themselves. There is no spin–diffusion in the case of carbon relaxation and the magnetization transfer here can be realized only by the transfer of the defect itself. To explain the effects connected with the cross-polarization we would like to assume that the defects are characterized by their own cross-relaxation times (T_{CH}). So, defects with large amplitudes, which have large relaxation rates (T_{CH}), do not contribute to the ^{13}C magnetization and we detect a signal determined mainly by the small defects. Since the cross-relaxation rates of the small defects are not large with regard to the diffusional rates, so the observed signal is averaged by the diffusion process. Thus, the mechanism of the defect diffusion can explain the narrower homogeneous spectra and the smaller q^2 values observed for the β_{cr}-transition in comparison with those for the β_{am}-transition in PE.

Hence, all experimental data of the relaxation of ^{13}C and ^1H in the region of the $\beta_{cr,\,int}$-transition can be coordinated within the limits of the concept of torsional oscillations using the model of defect diffusion. The fact that q^2 for this transition grows if we proceed to the intermediate phase (i.e. to more imperfect or smaller crystals) does not contradict the offered interpretations. The growth of the activation energy for the $\beta_{cr,\,int}$-transitions in comparison with the β_{am}-transition is connected with the fact that the formation of the free volume necessary to re-orient several CH_2-groups in the crystal needs more energy than in the amorphous sample.

In the same way one can also understand the mechanism of the γ_{am}-transition, which is most easily observed in branched PE [263]. Here the fast rotating side- and end-CH_3-groups represent the defects. These groups create the strongest relaxation mechanism at low temperatures, while the averaging process is realized by spin-diffusion. The width of correlation time spectra (≈ 0.3) as well as the sharp decrease of the activation energy ($\approx 13\,\text{kJ/mol}$) support the facts mentioned. Generally speaking, the existence of the end-CH_3-groups in polymers is the main hindrance to the study of the oscillation motions of the backbone by nuclear magnetic relaxation of protons. In crystalline regions of PE the oscillations of the methylene groups can be fixed and identified, that is due to the fact that during the crystallization process in LPE the methyl groups which represent the rare defects, are forced out to amorphous regions and are practically absent in crystallites, while the amorphous regions are being enriched by these groups. The α-relaxation transitions in the amorphous regions of PE and even in the defective PE crystals can be explained within the limits of our concept by the fact that in the temperature region of the α-transitions the conditions are created for the appearance of a great number of defects with great dimensions and amplitudes. This circumstance sharply increases the efficiency of the averaging of the dipole–dipolar interaction and creates a seeming isotropy of the rotation. The great width of the correlation time spectra observed in the region of the α-transitions ($\beta \approx 0.2$ to 0.3) supports this point of view, which is also characteristic for purely amorphous polymers.

In summary it should be noted that the total set of the microdynamic parameters obtained from NMR data and describing eight different relaxation transitions in PE does not contradict the concept of the accumulation of the torsional oscillations for describing the mechanisms of the molecular motion of the PE polymer chain. Thus, for five transitions ($\gamma_{cr,\,int}$, β_{am}, $\alpha_{am,\,int}$) this concept completely describes the behaviour of all parameters, while for the rest the complete description is made using the model of defects because the mechanism of the torsional oscillations used for all CH_2-groups would be too strong.

4.2.4 Connection between Molecular Motion and Structure in PE

The results of the investigation of a series of PE samples of various molecular mass and branching which were carried out by a structural (WAXS) and a dynamic (NMR) method will be compared and discussed in this section. The investigations of analogous PE samples were made in two laboratories by means of the NMR-pulse method [264] and by the WAXS method [265] simultaneously. In both cases the interpretation of the data was done within the limits of the three-phase model of polymer structure. All this provided the best comparison of structural and dynamic parameters and made it possible:

(i)   to observe the influence of the PE molecular structure on its phase structure,
(ii) to test the correspondence of the data of the two methods and
(iii) to touch upon a question of correlation between the structure parameters and
      the characteristics of the relaxation transitions.

In NMR experiments the TMD shape was studied for a wide temperature range. The procedure described in detail in Chap. 3 was used to analyse this shape.

Table 11. Parameters of the relaxation transition [264] and of the structure [265] in several PE samples $\Delta(2\theta_{av}^{int})$ – differences of the X-ray densities, $\Delta\chi$ – differences of the angles of chain orientation of the intermediate and crystalline phases. 1–6: LPE, differences taken for 378 and 288 K, 7: BPE, differences taken for 341 and 288 K. All temperatures in K, activation energies in kJ mol^{-1}. Mw/Mn $\approx$ 2.5 for samples num. 1, 3, 4, 5, 6 and $\approx$ 8 for num. 2, 7.

Sample Num.	M_n	T_α^{am}	T_α^{int}	T_β^{am}	T_β^{int}	T_β^{cr}	T^*	$E'_{am}(\beta')$	$E''_{am}(\beta'')$	$E_{int}(\beta)$	$\Delta(2\theta_{av}^{int})$	$\Delta\chi$	$2\theta_{av}^{int}$
1	5×10^4	283	330	170	283	373	—	4.3(0.16)	—	6.4(0.23)	0.7	13	21.2
2	1×10^5	288	339	170	275	353	349	4.9(0.17)	13(0.41)	9.0(0.30)	1.1	18	21.1
3	3×10^5	293	393	170	301	371	373	3.6(0.13)	12(0.39)	—	—	—	—
4	4×10^5	290	330	—	—	—	376	5.5(0.19)	17(0.50)	8.5(0.29)	—	—	—
5	8×10^5	258	317	—	—	—	380	3.5(0.13)	—	5.7(0.21)	—	—	—
6	3×10^6	256	284	170	263	373	—	5.1(0.20)	—	5.5(0.20)	0.7	0	20.8
7	5×10^4	250	270	150	—	—	—	5.3(0.20)	—	4.4(0.16)	0.9	0	—

The analysis of the WAXS data was made according to the procedure described in [265]. The main polymer characteristics are given in Table 11.

4.2.4.1 Phase Structure

Figure 65 presents the dependences of the parameters P_{cr}, P_{am}, P_{int} on M_w. It can clearly be seen that the intensity of the crystalline component increases slightly with the growth of M_w to 3×10^5, then sharply decreases to 0.53 at 4.3×10^5 and then remains constant. However, the intensity of the amorphous component has its minimum value at $M_w = 3 \times 10^5$ while it reaches its maximum at high M_w. P_{int} does not vary with the growth of M_w up to 8×10^5, and then increases by 10%. In branched PE P_{cr} is much lower (39%) and P_{int} is much higher (42%) than in linear PE.

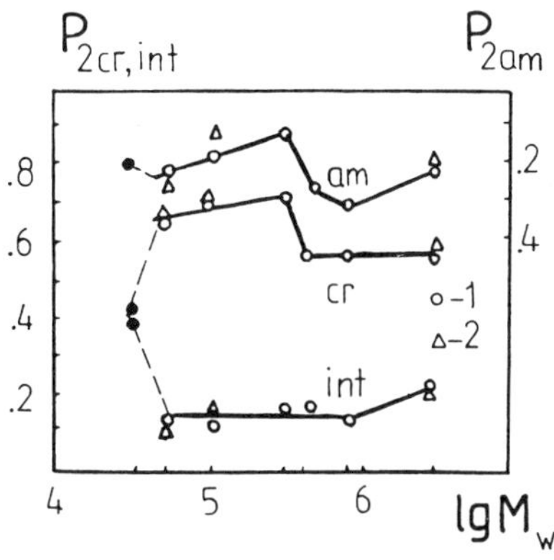

Fig. 65. Dependence of the intensities of the FID components (*1*) and of the WAXS components (*2*) on the molecular masses.

The intensities of the components of the X-ray scattering signal obtained for the analogous PE samples in [265] are also presented in Fig. 65. Since there is a dependence of the intensities on the temperature which is connected with the distribution of the signal between crystalline and intermediate phases as it is shown in this paper, we shall use the data obtained at certain temperatures (105 °C for linear PE samples and 68 °C for branched PE). One can see from Fig. 65 that the results of both methods are in good agreement with each other. This means both methods contain practically identical information on the PE phase structure. It should be noted that some differences in the values of P_i obtained by different methods can be caused by the fact that in [265] oriented but in [264] isotropic samples were studied.

Thus, it is stated by both methods that the phase structure of PE can properly be described within the limits of the three-phase model and depends on the molecular structure (M_w and branching), which makes it possible to regulate the molecular structure in a corresponding way and to prepare samples with different phase structures.

4.2.4.2 Molecular Motion

Let us consider the process connected with the development of the liquid-like motions within a certain phase. This process can be detected as the α-relaxation transition of one or another non-crystalline phase by the variation of the shape of the corresponding signal component from Gaussian to exponential. The beginning of the decrease of the second moment which is not accompanied by the variation of the signal shape, is connected with the β-relaxation process, which is caused by the development of the local anisotropic motions of the sections of backbone of the given phase.

Figure 66 shows the temperature dependence of the second moments of each phase at temperatures lower than the corresponding α-transitions, while Fig. 67 contains the temperature dependences of the spin–spin relaxation times of the amorphous and intermediate phases for temperatures higher than those of the α-transitions in all investigated PE samples. These temperatures are indicated by arrows in Fig. 66.

Figure 66 shows that the temperature dependences of the second moments of the crystalline and amorphous phases $\langle\Delta\omega^2\rangle_{cr}$ and $\langle\Delta\omega^2\rangle_{am}$ and correspondingly the temperatures of the β-transitions within these phases are approximately equal for LPE samples. As to the intermediate phase the temperature of the β-transition and the course of the temperature dependence at temperatures higher than T_β^{int} vary from one sample to another. It should be noted that in the amorphous phase the second moment at high temperatures reaches a plateau with the value of $\approx 3.5 \times 10^9\,\mathrm{s}^{-2}$, while in the intermediate and crystalline phases the decrease of the second moments proceeds to $7\text{–}8.5 \times 10^9\,\mathrm{s}^{-2}$ and then breaks off by the beginning of the more general α-process.

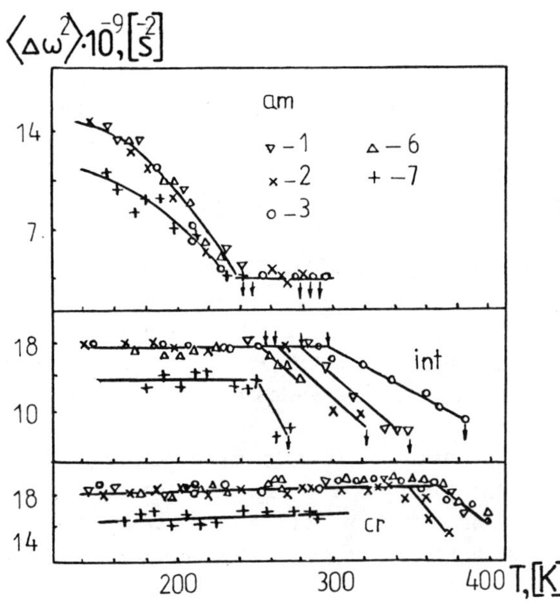

Fig. 66. Temperature dependence of the second moments in the amorphous, intermediate, and crystalline phases of PE. Parameters on the curves: sample number according to Table 11

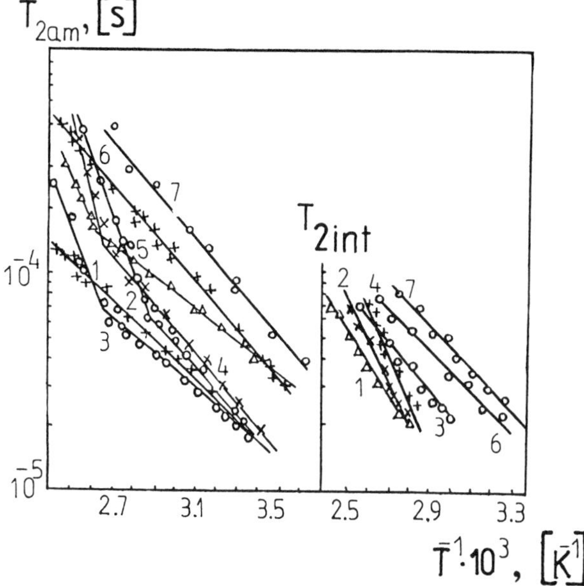

Fig. 67. Temperature dependence of the transversal relaxation time in the amorphous and intermediate phases of PE. Parameters on the curves: cf. Fig. 66

The temperature dependences of the second moments for each of the BPE phases are analogous to the corresponding dependences observed for LPE, but are shifted to the region of low temperatures.

The temperature dependences of the relaxation times T_{2am} and T_{2int} presented in Fig. 68 are properly described by straight lines in an Arrhenius representation, although for the samples 2 to 5 (cf. Table 11) there is a sharp variation of the slope of this dependence. The apparent activation energies (E_{am} and E_{int}) are determined

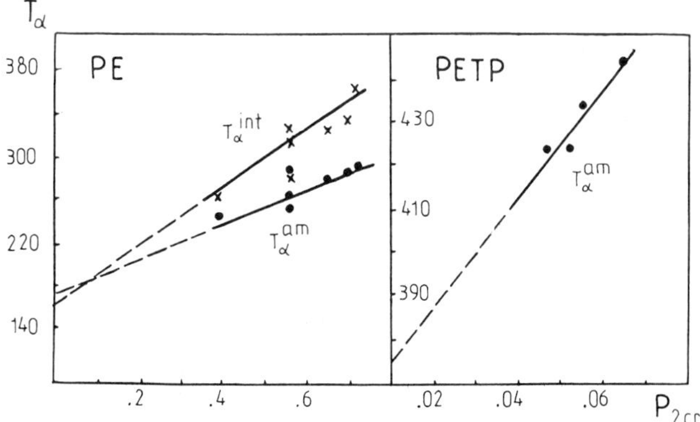

Fig. 68. Dependence of the α-transition temperatures in the amorphous (PE, PETP) and intermediate (PE) phases on the intensity of the crystalline phase

by means of the temperature dependences of the relaxation times, and for each of the samples 2–5 E_a values are obtained which correspond to the low- (E_a') and high-(E_a'') temperature sections of the dependences $T_{2am}(1/T)$. Besides this the temperatures of the turning-point (T^*) of these dependences are determined.

According to the formalism developed in [66, 150], width parameters of the spectrum of the correlation times are determined by the ratio of the apparent activation energy to the true value E^0 by the formula $\beta = 2E/(E^0 + E)$ (because Eqs. (78, 80) give with $\alpha = \beta$ for $\tau \ll T_2 : T_2 \sim \tau^{-\beta/(2-\beta)}$ instead of e.g. $T_1 \sim \tau^{-\beta}$). The spectrum of correlation times is assumed to be determined by a symmetrical ($\alpha = \beta$) function (cf. Eq. (80)), and the true value of E_a of about 210 kJ/mol to be identical for α-transitions in the amorphous and intermediate phases of all polymers.

The quantities of activation energies, the temperatures of α- and β-transitions, the parameters of the width of the spectra of the correlation times determined for seven PE samples, are presented in a table together with some data on the structure obtained by X-ray scattering in [265].

The main peculiarities of the data obtained are the following. The temperatures of the β-transitions in the crystalline and amorphous phases do not depend on M_w but they do so strongly in the intermediate phase.

There is a turning-point on the curves of the temperature dependences of T_2 in the samples with medium molecular masses. The temperatures of the turning-points coincide with the temperature of the β-transition in the crystalline phase.

The quantities T_α^{am}, T_α^{int}, and E_{int} reach the maximum values for samples with medium molecular masses (3×10^5), while the quantities T_β^{am}, T_β^{cr} and E_{am} are independent of M_w.

4.2.4.3 Comparison of the Structural and Dynamic Parameters

As we reached the experimental data by the two methods with the same samples we are going to establish some connections between the two aspects of the PE-structure — the static and dynamic one. Thus we shall restrict ourselves to the examination of the most evident correlations between the parameters of the structure and of the molecular motion.

From the comparison of the data of the two methods given in Table 11 and in Figs. 65, 67 we can conclude the following:

(i) The temperature of β-transitions of the noncrystalline phases correlates with their density determined by the X-ray method, the quantity T_β^{am} as well as the X-ray density do not depend on M_w and the quantities T_β^{int} and $(2\theta_{max})_{int}$ are connected with each other by a linear dependence.

(ii) The temperatures of the α-transitions of the amorphous and intermediate phases do not correlate with their density but with their content, i.e. with the phase structure. It seems that these temperatures are connected by a linear dependence with the intensity of the crystalline phase and do not correlate with the common intensity of the crystalline and intermediate phases. The extrapolation of both linear dependences $T_\alpha^{am,\,int}(P_{cr})$ to the zero crystallinity leads to a temperature of -90 to $-100\,°\mathrm{C}$. According to the physical meaning of such an extrapolation this temperature can be taken as T_g of the hypothetic amorphous PE (Fig. 68). This conclusion is affirmed by the analogous dependence we found studying various samples of PETP [38]. From Fig. 68 it can be seen that the dependence of T_α^{am} on P_{cr} in PETP is properly approximated by a straight line, which at $P_{cr} = 0$ gives a temperature close to T_g of the purely amorphous PETP (360 K).

(iii) The extreme behaviour of the temperature dependence of the relaxation times (the existence of a turning-point for the samples with medium M_w) and of some of their parameters (E_{int}, T_α^{int}) correlates with the extreme dependences on M_w of such structural parameters as temperature variations of X-ray density and relative disorientations of the chains of the intermediate phase. Though the model of the PE structure having been suggested in [265] explains the extreme behaviour of the structural parameters rather well, we think that at present there are not enough foundations to correctly describe the connections between the static and dynamic aspects of the structure established above within the limits of any molecular model.

The equality of the temperature T_β^{cr} and the turning-point temperature might be explained by the fact that the local motions developing at a given temperature in the crystalline phase eliminate the limitations of the mobility of some chains of the amorphous phase (e.g. the transit molecules or free ends), which in turn leads to a narrowing of the spectrum of the correlation times of the segmental motion, and, as a consequence, to the increase of the activation energy of this motion. The approximate equality of the temperatures T_α^{am} and T_β^{int} indicates the mutual connection of the local motions in the intermediate and of the large-scale motions in the amorphous phase.

5 Special Polymer Systems and Phenomena

5.1 Relaxation Times and Mechanical Properties

In their basic paper [200] concerning the ^{13}C-high resolution in solid glassy polymers Schaefer et al. discuss a connection between mechanical properties, especially the impact strength, in glassy polymers and ^{13}C relaxation parameters. On the other hand there are insufficiencies when mechanical properties like toughness and impact strength are studied by low-temperature mechanical or dielectric loss spectra. Therefore they propose characterizing the molecular origin of main chain motions by spectral densities which describe the main chain motion of a polymer at a single temperature, and to compare this spectral density with that of the impact itself. This spectral density of the polymer chain can roughly be determined by the ratio of the relaxation times $T_{\mathrm{CH}}/T_{1\varrho}^{\mathrm{C}}$. The arguments for this selection are firstly that an impact with a typical duration of about 100 μs shows a spectrum from a few kilohertz up to approx. 100 kHz, and secondly that the relaxation rates $(T_{1\varrho}^{\mathrm{C}})^{-1}$ and $(T_{\mathrm{CH}})^{-1}$ characterize the spectral intensities of the main chain at frequencies of about 30 kHz and near zero, respectively. That means their ratio, mentioned above, is a measure for the spectral density of the main chain motions in the mid-kilohertz range in relation to that for near static components. If, in a polymer, there are such intensive motions at the frequencies characterizing the impact, then these motions can initiate the cooperative nonlinear plastic flow, which causes the energy dissipation to the impacts and consequently increases the impact strength. Qualitatively it can be assumed that the spectral density of a mechanical impact is comparable with that of more flexible glassy polymers. However, rigid glassy polymers have more components at lower frequencies and not enough in the mid-kilohertz region which is also true for rubbers above the glass transition temperature, which have more high-frequency components. Therefore in the two latter cases there is a mismatch in the mid-frequency range and the polymer is unable to respond quickly to the impact, i.e. the impact strength is reduced in comparison with more flexible glassy polymers.

These general predictions were verified by investigation of seven glassy polymers. The plot $T_{\mathrm{CH}}/T_{1\varrho}^{\mathrm{C}}$ against the impact strength [200] (see [333] also) clearly shows a monotonous growth of both values starting with poly(methyl methacrylate) and finishing with polycarbonate. That means that this method in a physically well founded manner establishes the connection between pure NMR relaxation parameters and mechanical properties.

The relaxation time T_{CH} used here is that for the cross-polarization by adiabatic demagnetization technique in the rotating frame not described here. This time is similar to the cross-relaxation time T_{CH} discussed in Chapters 1 and 2 and has to be reduced by removing its dependence on the proton–proton dipolar interactions. It seems to be a good modification to substitute that time by the carbon spin–spin relaxation time T_2, sometimes measured in recent years [293, 225] and also reflecting the static interactions. Moreover, in [200] it is proposed that to describe the spectral density more accurately, some $T_{1\varrho}^{C}$-values at various frequencies be used. In other papers the applicability of this method is also discussed. For instance in [295, 296] it is shown that a general correlation between microscopic motions in modified polycarbonates and the impact strength can be established only in thin films, whereas the correlation fails for thicker specimens. In [297, 298] for polyethylene terephthalate and polyethylene the annealing influence is investigated, which is usually accompanied by a decrease in modulus.

5.2 Translational Motion of Macromolecules

Pulse NMR provides the unique possibility of studying translational motion in very viscous media. That means that the self-diffusion coefficient (D_s) can be measured with the pulsed magnetic field gradient technique within a range of 10^{-9} and $10^{-14}\,\mathrm{m^2\,s^{-1}}$ [299].

The investigations of the dependences of self-diffusion coefficients on the concentration and molecular mass are of great importance for testing models of the motion of macromolecules. Thus it is expected within the reptation model [300, 301] that in entangled melts and solutions,

$$D_s \sim M^{-2} \tag{181}$$

holds. In semidilute solutions the concentration dependence is described by the expression

$$D_s \sim c^{-\alpha}, \tag{182}$$

where α varies from 0 to 3 when turning from the diluted to the beginning of the concentrated region [300]. Recently a number of papers and reviews on self-diffusion in polymer systems has appeared [302–304].

The primary information in the experiments is the transversal magnetization decay $A(g^2)$ as a function of the pulsed magnetic field gradient. $A(g^2)$ is exponential for low-molecular liquids in the form of

$$A(g^2) = \exp(-\gamma^2 \delta^2 g^2 D_s t_d), \tag{183}$$

where g is the magnitude, δ the duration of the field gradients and t_d the time of diffusion observation.

Often the shape of diffusional magnetization decay in polymer melts and solutions is nonexponential and can be described by the expression [304, 305]

$$A(g^2) = \sum_i P_i \exp(-\gamma^2 \delta^2 g^2 D_{s_i} t_d). \tag{184}$$

Detailed investigations of the magnetization decay $A(g^2)$ in melts of low-dispersed PEO samples [306] have shown that $A(g^2)$ was exponential for samples with

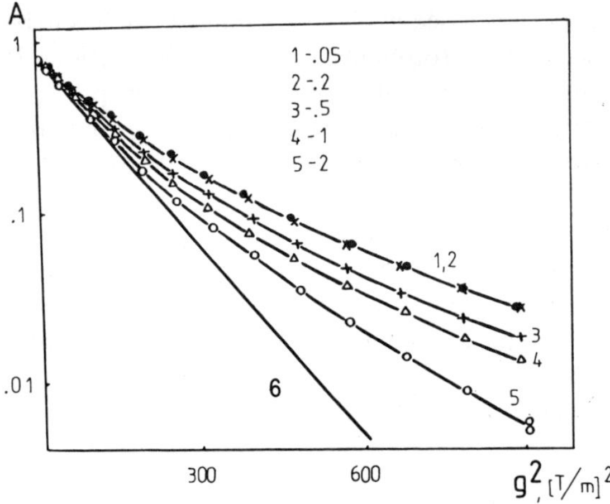

Fig. 69. Typical decays of diffusion damping in PEO with $M_w = 2 \times 10^3$ for different $t_d (\gamma^2 g^2 \delta^2 t_d = \text{const})$. Parameters: t_d in s, curve 6: initial slope

$M_n \lesssim 10^3$ whereas it could be described by Eq. (184) for $M_n > 10^3$. It is interesting to note that the shape of $A(g^2)$ depends on the diffusion time t_d with the tendency to become more exponential with increasing t_d (cf. Fig. 69). From the same figure one can see that the mean self-diffusion coefficient $D_s = \Sigma P_i D_{s_i}$ determined from the initial slope is independent of the diffusion time t_d. From the investigation of a great number of various polymer systems the authors conclude [303] that the nonexponentiality of $A(g^2)$ depends on the concentration of solutions, molecular mass, solvent and temperature. In [305], there is a hypothesis explaining these results according to which the translational mobility of macromolecules appears to be fluctuating in time and space because of the stochastic character of the formation of the so-called entangled network between the macromolecules. According to these explanations the nonexponentiality of $A(g^2)$ on the one hand provides the information on the fluctuations of translational motion and on the other requires special care in choosing the parameter characterizing the translation of the macromolecules adequately. The authors believe that only the mean self-diffusion coefficient determined from the initial slope of the decay can be such a parameter since it depends neither on the frequency of fluctuations nor on the choice of the observation time t_d.

Investigating the dependences of D_s on the molecular mass the dependence $D_s \sim M^{-2}$ was found for melts of polyethylene and polystyrene [307, 308] and poly(ethyleneglycol) [306]. No irregularity at M_c was seen. By measuring the diffusion of labeled polymer molecules in different matrices using optical and scattering methods it could be shown that in nonentangled systems the diffusion is of the Rouse type $(D \sim M^{-1})$ [309]. In the NMR self-diffusion measurements free volume effects are present for molar masses $M \lesssim M_c$.

In semidilute polymer solutions the validity of Eqs. (181) and (182) was observed by many authors [310–312] confirming the blob-concept of polymer solutions and the reptation mechanism. At high concentrations polymer–polymer friction leads to a very strong concentration dependence of the self-diffusion coefficient (cf. Fig. 70) [313].

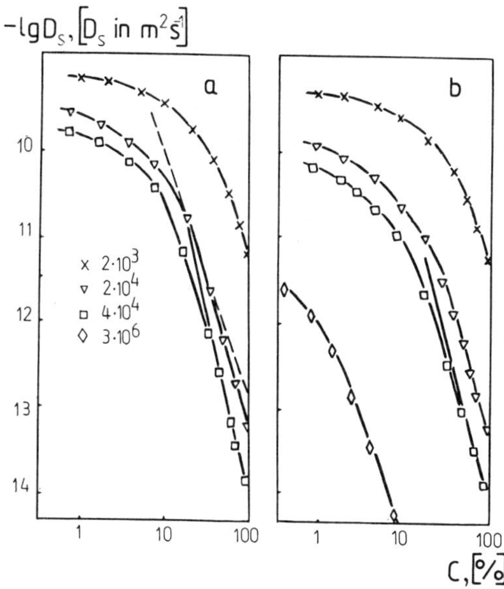

Fig. 70. Concentration dependence of the self-diffusion coefficient for PEO in solutions of (a) benzene and (b) dioxane. Parameters on the curves: molecular masses PEO

In [314], Maklakov and coworkers analyzed a great number of experimental data and showed that all the data on the concentration dependences of D_s are independent of the polymer type, of solvent quality, temperature and molecular mass, and are well described by the general scaling dependence:

$$D_s(c) = D_0 \cdot T_{2m}(c)/T_{2m}(0) \cdot f(c/c^*), \tag{185}$$

where D_0 is the self-diffusion coefficient if c tends to zero, $T_{2m}(c)$ and $T_{2m}(0)$ are the polymer spin–spin relaxation times at given and zero concentrations, $f(c/c^*)$ is a

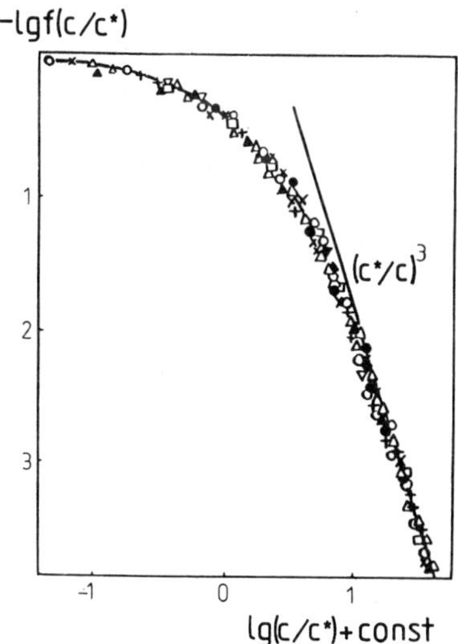

Fig. 71. Concentration dependence of the universal function f according to Eq. (185) for PEO and PS solved in chloroform ($\bigcirc$), benzene ($\times$), tetrachlorcarbon ($\triangle$), and water ($\blacktriangle$) for different molecular masses and temperatures

universal function of the dimensionless parameter c/c^*, c^* is the critical concentration that depends on the polymer type, molecular mass, solvent quality etc. In Eq. (185) the factor $T_{2m}(c)/T_{2m}(0)$ takes into account the variations of the small scale mobility of the polymer chain with the variations of concentration c, i.e. all the specific properties of the concrete macromolecule. The function $f(c/c^*)$ only reflects the most general universal peculiarities of the dynamics of the long polymer chains that are usually considered by theory. Figure 71 shows the dependence of $f(c/c^*)$ on c/c^*. It can be seen that this function unlike from the initial dependence D_s on c is in good agreement with the theoretical predictions, i.e.

$$f(c/c^*) \sim (c/c^*)^\alpha, \quad \text{where } 0 \le \alpha \le 3. \tag{186}$$

5.3   Multicomponent Polymer Systems

5.3.1   General Considerations

In recent years it has been shown that NMR is a powerful method of investigating polymer blends like simple mixtures, blends, block copolymers, filled polymers and polymers modified by several additives. The main reason for the production of such composites is the improvement of the mechanical and thermal properties or the use of cheaper materials without a substantial loss of properties essential for the applications desired. Although the desired properties mainly depend on the morphology of the composites under consideration and NMR generally provides only qualitative conclusions regarding morphological properties, several NMR techniques yield important information about these heterogeneous systems. By means of the investigation of the molecular motion NMR can e.g. estimate the compatibility in polymer blends, the mobility of the components and their spatial extent and the composition of the measured regions including intermediate regions which might arise.

The possibilities of these investigations, of course, depend on the special NMR parameters of the components in the polymer compound. They require e.g. differences in the relaxtion times or chemical shifts of the homopolymers forming the blend. Certain circumstances can favourably influence the interpretation, e.g. the presence of side-groups with intensive motions and fast relaxation effective as relaxation sinks, the use of deuterated substances, or the presence of fluorine nuclei in the samples, which permits the use of cross-relaxation measurements.

5.3.2   Some Methodical Aspects

5.3.2.1 Proton Measurements

Most of the proton and the ^{13}C experiments are based on methodical questions (mainly concerning spin dynamics, especially spin diffusion), which are discussed generally for heterogeneous systems in Chap. 3. Here we shall summarize some treatments typical for multicomponent polymer systems.

The starting point is the question whether an exponential relaxation behaviour exists. In the case of a loose mixture of two polymers without interactions affecting the NMR relaxation one measures a sum of two exponential functions only. The relaxation times as well as the intensities agree with the values for the homopolymers or with their relative proton content.

If there is a certain contact between the two components in dependence on the method used an averaging of the relaxation times can occur. The respective relaxation time T_1 is a limitation of the spin-diffusion process and connected by

$$\langle r^2 \rangle = 6DT_1 \tag{187}$$

with the mean square distance that the spin energy may cover during the relaxation time according to Eq. (41). With a spin-diffusion coefficient in the order of $10^{-12}\,\mathrm{cm}^2\,\mathrm{s}^{-1}$ this yields mean distances of approximately 25 nm and 0.8 nm, respectively for spin–lattice relaxation times T_1 of about 1 s and for spin–lattice relaxation times in the rotating frame $T_{1\varrho}$ of about 1 ms. If T_1 shows an effect of the spin-diffusion whereas $T_{1\varrho}$ does not, one can conclude that there are micro-heterogeneities with a linear extension between these two limits. That means that the time T_1 is sufficiently long so that the energy can arrive at the region of the other component, whereas $T_{1\varrho}$ is too short.

This conclusion concerning micro-heterogeneities was drawn by McBrierty et al. [315–318]. For PMMA/PSAN blends they found a segregation of the components on a scale between 2 and 15 nm. In the papers mentioned a more detailed analysis of the presence of effective relaxation sinks in both polymer components (e.g. methylene and phenyl side-groups) was made. When the mixing is complete, i.e. the spin-diffusion is effective, for the total measured relaxation rate $(T_l^t)^{-1}$ it follows

$$\frac{1}{T_l^t} = \frac{N_a}{N_t}\frac{1}{T_l^a} + \frac{N_b}{N_t}\frac{1}{T_l^b} \tag{188}$$

in dependence on the intrinsic rates $(T_l^a)^{-1}$ and $(T_l^b)^{-1}$ for the sinks with $N_{a,b}/N_t$ as the fraction of protons for the respective component acting as relaxation sinks.

Substituting N_a, N_b, N_t by the number of protons in a monomeric unit of each component, by the corresponding number of fast relaxing protons and by the weight fraction W_a of the component a, Eq. (188) yields a form where N_t/T_l^t as well as the sum $N_a/T_a + N_b/T_b$ depend linearly on W_a. Thus a plot of the former against W_1 will yield a straight line, if the spin-diffusion is dominant. Otherwise one gets a concave curve indicating that the spin-diffusion only has a partial or no effect, even if the signal is purely exponential. The conclusion mentioned above concerning the microheterogeneities in PMMA/PSAN blends was drawn from the fact that the $(N_t/T_l^t)/W_1$-plot shows a straight line for T_1 and a clear deviation from the linearity for $T_{1\varrho}$.

Another method important for multicomponent systems is the Goldman–Shen experiment, discussed in Chapters 1 and 3 [117]. Applications for polymer composites are demonstrated in [118, 316].

5.3.2.2 Carbon-13 Measurements

There are also some interesting possibilities for the application of the cross-polarisation/magic-angle spinning technique, all developed by Schaefer et al. Firstly we will refer to the remark made in Chap. 1: that the selective measurement of the magnetization of several carbons polarized in different ways by the protons provides a possibility of determining the proton relaxation time in the rotating frame $T_{1\varrho}$–^{1}H selectively, whereas direct proton measurements only show a mean value owing to spin-diffusion averaging. Stejskal et al. [321] have applied this technique to determine proton $T_{1\varrho}$ values for both components of a PPO/PS blend. The results exhibit essentially the same values of $T_{1\varrho}$ of protons for all carbons, indicating locally well blended chains. In contrast, a blend of PPO with poly(p-chlorostyrene) shows similarities in the $T_{1\varrho}$–^{1}H values with those in the homopolymers, which indicates the absence of any significant interchain mixing, even in the quenched material. Furthermore the paper discusses the proton–proton spin diffusion in a two-component system and, besides the initial relaxation rates for both components, the cross-relaxation rate is determined. In another paper, Schaefer et al. [322] have applied the same technique to an analysis of the interface between phases in a blend of perdeuteropolystyrene with a PS/PB block copolymer. In this case the $T_{1\varrho}$–^{13}C quantities which are shown to be sensitive to spatially dependent proton–proton spin diffusion are a measure of the proximity of protonated PS chains to one another. Additionally, another advantage of deuteration is used: Intermolecular cross-polarization of the protonated chain affects not only the intermolecular carbons but also the carbons in a deuterated chain. This provides further information concerning the mixing of protonated and deuterated chains. Both methods can be combined and for a deuterated polystyrene/K-resin blend lead to the conclusion that the interfacial region shows no structural features but rather consists of randomly mixed protonated and deuterated polystyrene chains.

5.3.3 Other Applications

Again we will not give a review of this extraordinarily wide field, but after principal remarks with methodical orientations, we will discuss some interesting examples.

5.3.3.1 Polymer Blends

Kwei et al. [323, 324] investigated compatible polymer mixtures using proton NMR T_1- and T_2-measurements besides other methods. The results for compositions of PS/PVME show a good compatibility of the films. However, especially T_2-measurements show the incomplete mixing on the time scale. The splitting of T_2 in a wide temperature region into two components was interpreted as the two phases according to the homopolymers, but each one extended by an interpenetrating region. Furthermore, thermally induced phase separation is studied and leads to a possible explanation of this separation by a spinodal mechanism.

The transient Overhauser effect in blends of PVF_2 and PMMA is used in [325] for studying the compatibility in PVF_2/PMMA and PVF_2/PEMA blends. This effect, discussed in the first chapter for the stationary case, can be used for the determination of an additional (besides T_1, $T_{1\varrho}$, T_2) relaxation parameter—σ/ϱ—characterizing the cross-polarization similar to the Overhauser-factor η. It was found that σ/ϱ approaches 0 at the temperature of liquid nitrogen in agreement with very small proton–fluorine cross-relaxation and characterizing a rigid lattice behaviour. The main result is that a substantial number of amorphous PVF_2 molecules have an immediate contact to those of PMMA and PEMA. We briefly mention that the cross-relaxation experiment enhances the sensitivity for T_1 components.

In some investigations in our laboratory [326, 327] PMMA/PVAc blends were studied using proton T_1 and $T_{1\varrho}$ and $^{13}C–T_{1\varrho}$ measurements. The results show a clear dependence on the history of the blends. As discussed with respect to the side-chain motions in PMMA, the α-CH_3-group also influences all other protons in the blends. This permits a treatment analogous to that described in Sect. 5.3.2.1 for the relaxation sinks. However, in this case the longitudinal relaxation function consists of two components, changed somewhat in comparison with the homopolymers. Therefore the procedure mentioned above is performed for the more modified PMMA component and results in a straight line. (The PVAC phase remains relatively undisturbed.) That means a good compatibility of the components in the mixture phase the linear dimensions of which can be estimated to be 15 nm. The reasons for the heterogeneities of this blend, usually considered to be well compatible, can be connected with the precipitation process during the sample preparation.

5.3.3.2 Block-Copolymers

Here we shall only discuss the diblock-copolymer PEO/PMMA [224, 328, 329]. The temperature dependence of $T_{1\varrho}$ for protons and for ^{13}C is measured for the copolymer in comparison with some results for the homopolymers. The ^{13}C measurements allow us to separate the signals of the α-methyl group and the quaternary carbon in the PMMA component, and of the methylene groups in the PEO part, although magic angle rotation was not applied. The facts of interest are the shifts of the relaxation temperature regions in comparison to the homopolymers. Firstly, the dynamic glass-transitions in PMMA, i.e. the minimum position of $T_{1\varrho}$–^{13}C is shifted by about 45 K to lower temperatures as a consequence of the influence of the PEO melt. Secondly, and contrarily, the α-CH_3-rotation is shifted by about 35 K to higher temperatures, caused by a motional hindrance through the inflexible chains in the PEO crystalline regions. The proton $T_{1\varrho}$ measurements support the results of the differential scanning calorimetry with respect to the melting and crystallisation behaviour and show that the sharp jump at a temperature of 330 K disappears by increasing the PMMA content. Indications of the emergence of an intermediate region could not be found.

5.3.3.3 Filled Polymers

For the investigation of polymer interactions with fillers the NMR also yields some possibilities. So one can use perhaps the deuteron resonance to study the interactions with the electric field gradients in the boundary layers.

Resing et al. [206] discuss the possibilities for an improvement of the generally weak bonding at the polymer–filler interfaces. They demonstrate the ability of the ^{13}C NMR with CP/MAS technique e.g. to prove the covalent bonding between silicagel and methyl-vinyl-dichlorosilane realized to promote the adhesion by reactive pendant groups.

Preparatory investigations in our laboratory [330] show substantial influences on T_2 of protons for the system PMMA/suprasil, which indicate an emergence of structural heterogeneities. Contrarily, T_1 measurements of protons only show a small influence of the filling on the relaxation times, whereas the relative intensities of the two components change owing to changed filler concentrations. Aerosil-filled poly(dimethyl siloxane) was studied by Litvinov [331, 343] using proton T_1 measurements. Besides the two minima corresponding to the methyl rotation and the glass transition which are observed in the pure polymer, an additional high-temperature minimum appears for increasing the filler content. This permits us to distinguish between polymer layers adsorbed at the filler and parts of the polymer not connected with the filler surface [286].

The use of TMD investigations seems to be more successful for a more detailed characterization of the filler influence. Some interesting results concerning the influence of fillers on the structure and dynamics of polymers were obtained investigating elastomers filled with carbon black. In [144, 147, 334] the appearance of a short component ($T_2 \approx$ 20–30 μs) in the TMD with an intensity of about 10 to 15% is shown, if active (fine dispersed) carbon is put into the rubber. That means, evidently, that a part of the macromolecules loses its segmental movableness.

An accurate quantitative analysis of the TMD shape in filled cross-linked rubbers (on the basis of PIP) [334] using the theory represented in Sect 3.5.1 yields information on the density of the chemical network in systems containing different concentrations of fillers with various dispersions.

Comparative investigations of the network density were done using mechanical standard methods and demonstrated in Table 12 [334]. In [147, 334] it was shown that the network parameters measured by both methods are in a rough agreement, if the polymers are not filled or only filled with inactive fillers. Table 12 shows that in the case of active fillers NMR measurements may give reasonable network densities even for systems where the modulus is much larger than for the unfilled systems.

Table 12. Network density in polyisoprene in dependence on the sulphur and the carbon-black content, measured by mechanical extension and by NMR (according to Eqs. (137, 139)) methods. E module of elasticity, index 0 means: without filler

Sulphur phr	Carbon-black phr	Mechanics E/E^0	NMR M_c^0/M_c
1, 0	50 (active)	4.3	0.83
3, 0	50 (active)	2.4	0.78
5, 0	50 (active)	1.6	0.86
1, 0	100 (active)	15.6	0.54
1, 0	100 (inactive)	2.8	1.01

However, in swollen polymers the results from both methods are approximately the same.

A comparative analysis of the results obtained by the two methods showed that the NMR method gives more correct quantitative information about the density of the vulcanization net. It was demonstrated that the bringing in of the filler has practically no influence on the density of the chemical network, although an active filler binds up to 15% of the macromolecules to its surface. Because the intensity of the component concerning the combined macromolecules tends to zero at high temperatures (≈ 470 K) but appears again during the following cooling down, one can assume that the bonding polymer-filler is not of a chemical nature. The effect of reinforcement found using deformation methods is evidently coupled with the formation of a dense physical network consisting of transition macromolecules uniting the neighbouring chains.

5.4 Molecular Motion in Globular Proteins

To study the dynamical properties of the proteins one can use both selective NMR methods (high-resolution spectra in solutions and solids) and non-selective ones (proton relaxation of a common spin-system in solutions, broad line spectra and relaxation in powders). In the first case the relaxation of separate lines is studied, a certain number of lines may be large enough and thus one can obtain information on the motion in various regions of the macromolecule. In the second case the information obtained is averaged over all macromolecular sites [335–339].

As to liquids and solutions of low viscosity, including solutions of synthetic polymers, a number of lines in the spectra observed usually correspond to a number of types of chemical groups, that is due to a rapid isotropic rotation of the small molecules or due to segmental motion of the polymer chain, which creates the optimal conditions for high-resolution. The high-resolution spectra in solids are usually broader, and therefore a smaller number of lines can be observed. It is because of the non-averaged chemical screening of the atomic groups, that results in the anisotropy of chemical shifts, and due to the smaller effectiveness of the dipolar interaction averaging by means of mechanical or radio-frequency pulse rotations of the sample. Sometimes the number of lines in solutions of native proteins is larger than that of denaturated proteins or of a corresponding amino-acid mixture, but the width of the lines is a little greater. This effect is explained in the following way: in native proteins the nuclear spins feel the screening not only from the electron shells of the nearest chemically tied neighbours but also from those of spatially closed groups which are at a great distance along the polypeptide chain.

It is only possible due to the lack of large scale fluctuations of polypeptide chains. Those fluctuations mainly average the interactions of the second type.

Thus, the situation is similar to that which occurs in the experiments on solids, where the high-resolution NMR spectra are observed by using mechanical rotation of a sample. But here the part of the solid is played by the protein macromolecule with a unique spatial structure, and its Brownian motion with frequencies of about 10^7–10^9 Hz plays the part of mechanical rotation.

To solve the question on the nature of the protein intermolecular motion it is necessary to study in detail the magnetic relaxation of separate lines in high-resolution spectra or of the common spin-system.

Though the latter only yields rough information averaged over all nuclei, it can, nevertheless, often be of some use and the only accessible information.

In this chapter we will consider two examples of the analysis of NMR relaxation experiments in proteins. The general aim of this experiment is to get information about mechanisms of the internal motions and their variations under the influence of conformational transitions.

5.4.1 Selective NMR Experiments in Solutions

Let us consider ^{13}C spectra because of the simplicity of their interpretation. At present the spectra which are resolved enough to carry out the correct measurement of single resonance lines, can be obtained for small proteins only. The best model for such experiments is one of the smallest proteins: bovine pancreatic trypsin inhibitor (BPTI). The high-resolution NMR spectra of this protein in D_2O consist of more than 30 well-resolved lines identified with various ^{13}C sites on the main and the side chains. Six independent parameters (T_1, T_2, NOE, all for resonance frequencies 45 and 90 MHz) are measured for every line [62].

This unique experiment due to its scope and its detailed elaboration is much better than any one that has been carried out so far by NMR or other dynamical methods. It yields the richest data set for analyzing the characteristics and mechanisms of intermolecular motion within globular proteins.

When one is going to get information from NMR data, the bottleneck is the transition from the experimental data to the microdynamical characteristics of the system. As has already been noted, a lot of authors recognize the traditional model approach to be unsatisfactory because this is more complicated. In the case of proteins it is not accompanied by improved exactness or by increased definiteness.

The experimental data mentioned above have been analyzed by means of three modifications of the model-free approach [62, 63, 339]. They were firstly described [62] with the suggestion that carbons take part in three independent motions being characterized by exponential correlation functions. In the next papers [62, 339] it was suggested that carbon intermolecular motion is anisotropic, so it was the first step towards the representation of a protein as a solid. It allows us to get a quantitative characteristic of spatial restrictions of intermolecular motions and to get nearer to a correct interpretation of the experiment. In spite of some different ways of thinking, all the authors came to the same conclusion: to describe the experimental data one can do so by considering just two types of motion. They are a local anisotropic motion and an overall tumbling of the macromolecule. In [290, 340–342] the next step of applying the model-free approach to the same experimental data was made. Here a suggestion was introduced that a local motion of the protein carbons can be described by the same laws as the atomic motion in some synthetic polymers, i.e. in general, the protein intermolecular motion is anisotropic and must be characterized by non-exponential correlation functions of Eq. (55).

In [340–342] the experimental data for BPTI were analyzed in detail assuming Eq. (189) for the correlation function for any carbon groups:

$$\phi(t) = A[(1-q^2)e^{-t/\tau_R} + q^2 \phi'(t/\tau_l)]. \tag{189}$$

It has been also assumed that the correlation time distribution function $\phi'(t/\tau_l)$ for the local motion has the form of Eqs. (56, 80). It is a symmetrical function ($\alpha=\beta$) and can be described by two parameters, β and τ_l.

The parameters q^2, τ_l, τ_R, and β were determined using a program of function minimizing by the polyhedron deformation method. The typical parameters are presented in Table 13.

Table 13. Typical microdynamical parameters for different chemical groups in BPTI

Group	α-CH	β-CH$_2$	γ-CH$_2$	$(\gamma$-$\varepsilon)$CH$_3$	ar-CH
q^2	0.30	0.65	0.98	1	0.40
β	0.50	0.60	0.60	0.65	0.40
τ_l[ps]	200	180	40	20	50

By analyzing the results one can yield the next regularities:

(i) The parameters of local motion anisotropy are raised in a row:

$$q^2_{\alpha\text{-CH}} < q^2_{\text{CH-ar}} < q^2_{\beta\text{-CH}_2} < q^2_{\gamma\text{-CH}_2} < q^2_{\text{CH}_3}.$$

(ii) The parameters of the width of the correlation time distribution are much smaller than 1 and decrease in a row:

$$\beta_{\text{CH}_3} \approx \beta_{\text{CH}_2} > \beta_{\alpha\text{-CH}} > \beta_{\text{CH-ar}}.$$

(iii) The effective correlation times of the local motion of α-carbons and of some β-carbons are of the order of hundreds of picoseconds, whereas for other groups this parameter is of the order of ten picoseconds.

The main feature of the results for any type of carbon groups is a great distribution of correlation times, which can be both of homogeneous and of inhomogeneous nature. The homogeneous distribution is a result of the complexity of the intermolecular motion and the inhomogeneous one is due to various microscopical environments of the same atoms in different macromolecules.

The possibility of the distribution to be of inhomogeneous nature is not evident in proteins. According to X-ray data there are no static inhomogeneities of the identical group environment in the macromolecular ensemble. So, one can deal only with a seeming dynamical heterogeneity of environments, which is not averaged during the time of 10^{-9} s.

Further analysis of the data, that is based on comparison of the obtained fitting parameters with those predicted by various models, is analogous to the analysis presented in detail in Chap. 4 for the synthetic polymers.

The short correlation times, the values of q^2 close to 1 and the high values of β are an evidence of the fact that the motion of CH_3- and CH_2-groups which are distant enough from the main chain can be considered as anisotropic hindered rotation. It is the result of summarizing of two processes, one of them being the group rotation about the symmetry axis and another—the restricted motion of this axis [57].

The aromatic aminoacid residue ring motion is characterized by small values of the anisotropy parameter ($q^2 \approx 0.4$), by small correlation times ($\tau_l \approx 50ps$) and by a broad spectrum of correlation times ($\beta \approx 0.4$), that is likely of inhomogeneous nature. Those data give an opportunity to assume that the ring local motion can be described by torsional oscillations [344, 345]. It we suggest that the torsional motion of the CH-vector is restricted by a cone with a half-angle δ we can estimate this δ by Eq. (90) for $q^2 = 0.4: \delta = 35°$.

Data for α-carbons (large τ_l, small q^2, $\beta \approx 0.5$) can be described both in the frame of the torsional oscillation model and by the mechanism of defect diffusion [87]. To distinguish these two cases one has to analyze the shape of nuclear magnetization decays. As to α-CH group relaxation there is a very large contribution of macromolecule rotation ($\approx 70\%$), which results in smoothing the effect of an inhomogeneous distribution of the correlation times of the local motion.

Therefore it is necessary to carry out a theoretical calculation of those decays for some concrete sets of microdynamical parameters before analyzing the experiment. In Fig. 72 there are presented the longitudinal and transverse magnetization decays calculated both for homogeneous and for inhomogeneous distributions according to Eqs. (74) and (75) with correlation function Eq. (189) and with

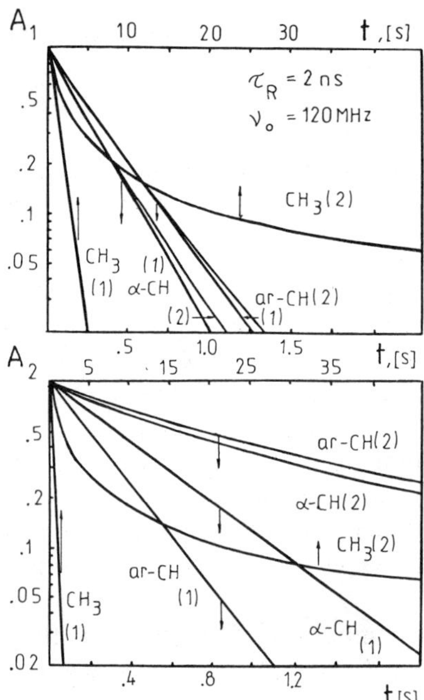

Fig. 72. Theoretical decay of the longitudinal and transversal magnetization, calculated by Eqs. (74, 75) using parameters of the correlation function Eq. (189) given in Table 13. In parentheses: 1-homogeneous, 2-inhomogeneous distribution

parameters presented in Table 13. From Fig. 72, one can see that, in fact, definite conditions exist which result in large deviations of the experimental signals $A(t)$ from exponential ones, so that deviations can be detected in the experiment. Till now, we have been unable to find in the literature any information on non-exponential α-CH magnetization decays. Moreover, the detailed analysis of α-CH group magnetization decays ($A_1(t)$ and $A_2(t)$), which was carried out at our request in the laboratory of Lippmaa [346], does not detect any essential deviation of signals (for bovine carbonic anhydrase in solution) from the exponential one.

Although, in our opinion, one should not make any definite conclusions without special experiments, nevertheless we think that the defect diffusion mechanism is preferable. Thus, the molecular interpretation of the motional mechanisms may be analogous to that we have used to analyze ^{13}C and ^{1}H relaxations in the crystalline phase of polyethylene at high temperatures (β-transition). According to this interpretation (cf. Chap. 4) a small number of α-CH groups oscillates with larger amplitudes than the others. Such groups play the role of defects. The probability of such excited states is not large. At the time of observation only a small number of all the chain atoms can suffer such excitation. So, the parameter q^2 will depend not only on the oscillation amplitude but also on the defect concentration. The parameter τ_0 will be connected with the time of defect diffusional displacement, i.e. with the migration rate of the excited state.

Thus, the experiments analyzed show that a macromolecule undergoes a Brownian rotation, and separate groups are also involved into one high-frequency rotational-oscillational motion of various amplitudes. So, the carbons of backbones and of aromatic rings take part in just restricted oscillations, but methyl and methylene groups distant from the chain rotate almost isotropically. It should be noted that no indication of a liquid-like motion of the protein peptide chain has been found in those experiments. Moreover, as to the backbone carbons, the same local motions characteristic of the main chain of glass-like polymers (α-transition) were not found. Analogous qualitative conclusion on strong restriction of molecular motion of polypeptide chain atoms was made in [345], where the traditional model approach to analyzing the data was used. High-resolution ^{1}H-NMR data, in spite of some difficulties in their interpretation (cf. e.g. [338]), allow us to reach the same conclusion about details of the protein dynamics in accordance with high-resolution ^{13}C-NMR data.

5.4.2 Non-Selective NMR Experiments

In [339, 347, 348] the mechanisms of the nuclear magnetic relaxation of protein protons in solid and liquid solutions of bovine pancreatic ribonuclease A (RNase) in heavy water (5% RNase in 99.6% D_2O, 0.1-molar NaCl, pD = 2.6) was studied to establish the mechanisms of the protein proton internal motion as well as a connection between that motion and the conformational state.

The following were analyzed: the shapes of transversal magnetization decays at temperatures 170 to 360 K (the FID in the range 170 to 270 K, and the transversal magnetization decays (TMD) by the method of Carr–Purcell–Meiboom–Gill in the range 276 to 360 K). The shape of the FID was described by a sum of two Gauss components:

$$A_2(t) = P_1 \exp\left(-\tfrac{1}{2}\langle \Delta\omega_1^2 \rangle t^2\right) + P_2 \exp\left(-\tfrac{1}{2}\langle \Delta\omega_2^2 \rangle t^2\right), \qquad (190)$$

and the TMD of protein protons can be presented well enough by a sum of two exponents:

$$A_2(t) = P_{21}\exp\left(-\frac{t}{T_{21}}\right) + P_{22}\exp\left(-\frac{t}{T_{22}}\right), \qquad (191)$$

where p_i are the intensities of the components, T_{2i} and $\langle \Delta\omega_i^2 \rangle$—the relaxation times and the second moments of corresponding components, and index 1 denotes the more slowly decaying component. At temperatures higher than 353 K the TMD transforms into a single exponent.

On first approximation one can consider that the slow component of the FID (which intensity rises from 0.08 to 0.17 with temperature increase) describes the proton relaxation of the most mobile methyl groups in the protein, whereas the fast decaying component describes the proton relaxation of methine, methylene and the rest of methyl groups.

The temperature dependences of the second moments and of the intensities of the slow FID components are presented in Fig. 73. The temperature dependences of the second moments, $\langle \Delta\omega_2^2 \rangle$ and $\langle \Delta\omega_1^2 \rangle$, are typical for solids with anisotropic internal motions: so, for $\langle \Delta\omega_2^2 \rangle$ the transition from the low-temperature plateau (170 to 200 K) to the high-temperature one (230 to 270 K) is observed, and for $\langle \Delta\omega_1^2 \rangle$ the high-temperature plateau (200 to 270 K) exists only. The identification of TMD components, which has been made for liquid solutions in [339], implies that at room temperatures the fast component (T_{22}) is caused by methylene group relaxation, whereas the slow component is due to the CH- and CH$_3$-group relaxation. The value of P_{22} is equal to the calculated number of the CH$_2$-group protons (0.46) and does not depend on the temperature in the range 276 to 310 K. At temperatures above 325 K this value diminishes and at $T \approx 353$ K it becomes zero (cf. Fig. 74). The temperature dependences of T_{22} and T_{21}, that are monotonous

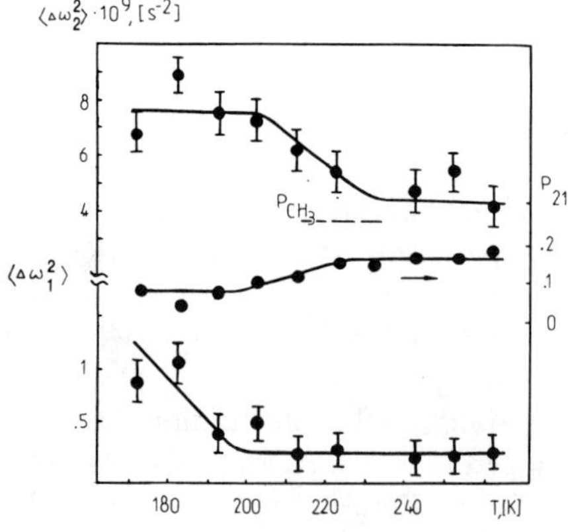

Fig. 73. Temperature dependence of the second moments for both and of the intensities for the slow components of the FID in a frozen solution of RNase

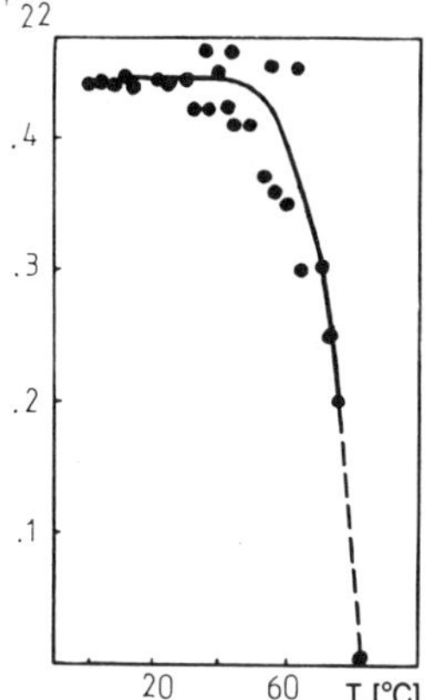

Fig. 74. Temperature dependence of the intensity of the short component of the transversal magnetization decay in RNase solution

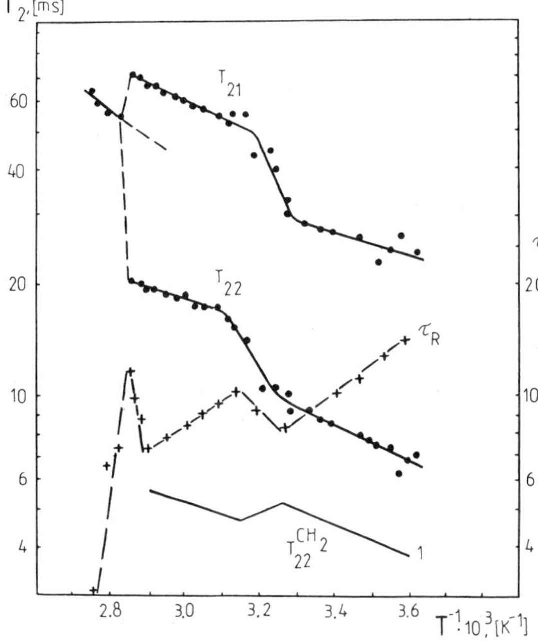

Fig. 75. Temperature dependence of the transversal proton relaxation times in proteins (T_{21}, T_{22}) and correlation times (τ_R) of the rotational motion of the macromolecule [289]. Solid lines: relaxation times calculated using parameters from Fig. 76, curve 1: $q^2 = 1$

in the ranges 276 to 308 K and 323 to 353 K and have the same slopes in Arrhenius coordinates, undergo a sharp jump at 315 and 353 K (Fig. 75).

At the first sharp increase of T_{2i}, which corresponds to the RNase reversible conformational transition, the relaxation time increases by a factor 1.5 to 2. At temperatures above 353 K the fast component disappears due to a sharp increasing of the relaxation time T_{22} which approaches to the time T_{21} and becomes distinguishless from the latter. These changes correspond to irreversible conformational transitions of RNase. Actually, the repeated measurement of TMD (after sample heating to 330 and to 370 K) results in a perfect reversibility in the first case and in an essential irreversibility in the second.

To analyze quantitatively the relaxation times and the second moments and to extract information from them on the molecular motion parameters, let us use the formalism which was developed in Chap. 1 and was applied to the analysis of ^{13}C relaxation data in PTI solutions. Let us assume

(i) The protein spin system consists of three weak-connected systems related to methine, methylene and methyl group protons. Each system is characterized by its own dipole–dipole interaction constant (by second moments equal respectively to $(3 \pm 1) \times 10^9$, $(12 \pm 1) \times 10^9$, $(14 \pm 1) \times 10^9 \, \text{s}^{-2}$, cf. e.g. [349]).

(ii) All intermolecular vectors take part in two types of motion: the intermolecular individual local anisotropic motion and the overall motion—the isotropic macromolecule rotation. The overall motion is described by a single correlation time function.

(iii) The spin–spin relaxation times can be described in general by

$$T_{2i}^{-1} = \langle \Delta\omega_{0i}^2 \rangle [q_i^2 f^i_i(\tau_l^i, \beta_i) + (1 - q_i^2) f_R(\tau_R)], \tag{192}$$

where τ_l^i, q_i^2 and β_i are the correlation time, the anisotropy factor and the width-parameter of the correlation time spectrum of the local motion, τ_R is the correlation time of the macromolecular Brownian rotation, and i means CH_3, CH_2 and CH.

At temperatures lower than 273 K Eq. (192) becomes

$$\langle \Delta\omega_i^2 \rangle = \langle \Delta\omega_{0i}^2 \rangle \left[(1 - q_i^2) + q_i^2 \frac{2}{\pi} \arctan \left(\frac{\pi}{2} \frac{\tau_l^i}{T_{2i}} \right)^{\beta_i} \right] \equiv T_{2i}^{-2}, \tag{193}$$

and at temperatures above 273 K ($\tau_l^i < 10^{-10}\,\text{s}$):

$$T_{2i}^{-1} = 1/3 \langle \Delta\omega_{0i}^2 \rangle (1 - q_i^2)[3J(0) + 5J(\omega_0 \tau_R) + 2J(2\omega_0 \tau_R)] \tag{194}$$

and

$$\tau_R = \tau_R^0 \exp \left(\frac{E_a}{RT} \right), \tag{195}$$

where τ_R^0 is the period of inertial oscillation of macromolecules inside a potential box (J according to Eq. (69)).

To determine the local motion parameters from the second moment temperature dependences let us use Eq. (194). Since the value of $\langle \Delta\omega_2^2 \rangle$ is mainly determined by the CH_2-group interactions, so the parameters extracted from the temperature dependences of $\langle \Delta\omega_2^2 \rangle$ will be characterized mainly by motions of those groups, i.e. $q_2^2 \approx q_{CH_2}^2$, $\tau_l^2 = \tau_l^{CH_2}$. Analogously let us assume that the value of $\langle \Delta\omega_1^2 \rangle$ involves an information on methyl group motions ($q_1^2 = q_{CH_3}^2$, $\tau_l^1 = \tau_l^{CH_3}$). To find the local motion correlation times for groups of two types we proposed $\beta_i = 0.5$. The results of calculations are presented in Table 14.

To describe quantitatively the high-temperature experiment let us use Eqs. (194, 195) suggesting that $E_a = 16$ kJ/mol in accordance with E_a of the viscous flow of this solution. The analysis of the experimental data showed that in the temperature

Table 14. Anisotropy parameters of the internal motion of RNase A

T [K]	200	230–270	273–315	325–353	353–370
CH_2	0.04	0.4	0.43	0.71	1
CH	0.04	—	0.3	0.65	1
CH_3	0.75(70%)	0.75(40%)	0.85	0.93	1
	0.98(30%)	0.98(60%)			

interval 275 to 350 K the behaviour of T_2 can be described by Eqs. (194, 195) with real behaviours of q^2 and of τ_R^0, but τ_R^0 must increase from 10^{-11} to 2×10^{-11} s in the range of the reversible denaturation transition. So, the result being a little unexpected, the direct measurements of τ_R in analogous samples were carried out at our laboratory (Kazan, Institute of Biology) using the time domain dielectric spectroscopy [288, 289]. The value τ_R actually rises at denaturation temperatures, τ_R^0 and E_a being equal to 1.1×10^{-11} s and 15.6 kJ/mol before denaturation transition and to 2.2×10^{-11} s and 15 kJ/mol after the transition. It almost coincides with the results of magnetic relaxation analysis.

We used τ_R of Fig. 75 for calculating the parameters q^2 for groups of all types. The results are presented in Fig. 76 and in Table 14. In Fig. 77 there are presented the correlation times of local motion of the CH_2- and CH_3-groups (τ_l^1 and τ_l^2) as well as the data of Table 13 for analogous groups of PTI. The latter were obtained by the analysis of ^{13}C relaxation data at room temperatures. It can be seen, that the correlation times of methylene and methyl groups form straight lines in Arrhenius coordinates with activation energies of 67 ± 8 and of 15 ± 3 kJ/mol, respectively.

All the results presented above can be explained if one assumes that, at temperatures here applied, protein can exist in three states: native (N), thermal denaturated (T) and unfolded random-coil conformation (U).

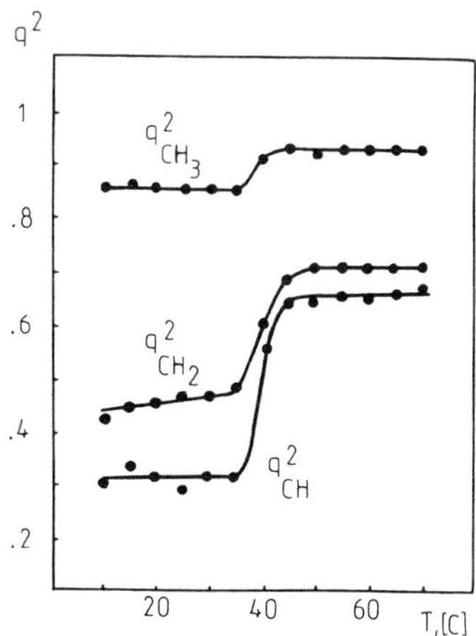

Fig. 76. Temperature dependence of the anisotropy parameter q^2 in RNase solution

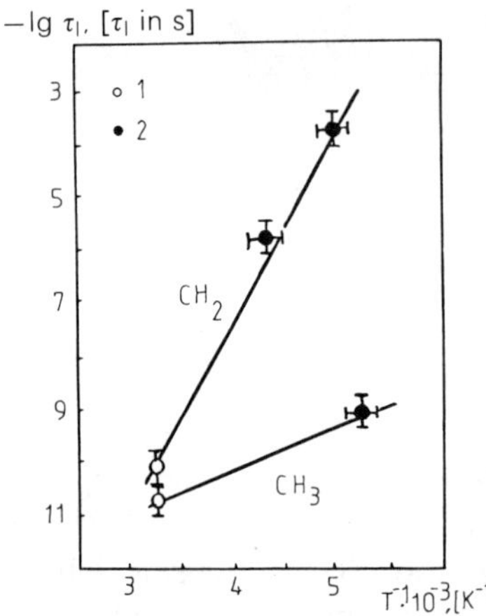

Fig. 77. Temperature dependence of the correlation time τ_0 of the internal motion in proteins for CH_2 and CH_3 groups. *1*: Data from Table 13 and [342]; *2*: from Fig. 73 using Eq. (193)

RNase exists in the native state at temperatures lower than 315 K, the most restricted intermolecular motion of all groups as well as the hard globule correlation time behaviour being correspondent to that state.

It is important to denote that during the transition from the solid to liquid solution the intermolecular motion amplitudes undergo no essential changes.

The thermal denaturated state is observed in the temperature range 325 to 353 K. It is characterized by a higher level of intermolecular motions with the conservation of globular structure elements, though some rising of τ_R compared with τ_R of the native state is observed. If one assumes that τ_R is described by the Debye relation (cf. e.g. [339, 350]) and hence τ_R is proportional to the volume of the macromolecule and also if one takes into account the change of the viscosity at the denaturation transition [350], so from the value of τ_R^N/τ_R^T of 1.5 to 2 (taken at some temperatures) one can conclude that the volume of the macromolecule in the denaturated state is twice as large as that in the native state.

The state of random-coil is observed at temperatures above the reversible denaturation transition, here, it exists together with the thermal denaturated state. The relative part of this state increases with the temperature increase. At temperatures above 353 K all macromolecules are in the state of random-coil. In that state the main mechanism of nuclear relaxation consists of segmental motions which are characteristic for solutions and melts of synthetic polymers (cf. Chap. 4).

It should be noted that, in experiments, two transitions between three states of RNase occur: the first—between states N and T—is a reversible phase transition of the first kind [351], it takes place in a narrow temperature interval $\Delta T \approx 10$ K and is followed by a globular volume increase and by a sharp decrease of the restrictions of the internal motions; the second transition occurs between states T and U in a rather wide temperature interval $\Delta T \approx 40$ K and is connected with a process of globule unfolding.

References

1. Abragam A (1961) The principles of nuclear magnetism, Clarendon, Oxford
2. Pake G E (1948) J. Chem. Phys. 16: 327
3. Andrew E R, Bersohn R (1950) J. Chem. Phys. 18: 159
4. Bersohn R, Gutowsky H S (1954) J. Chem. Phys. 22: 651
5. Engelsberg M, Lowe I J, Cardon J L (1973) Phys. Rev. B7: 924
6. Schmiedel H (1969) Phys. Lett. 29A: 682
7. Lundin A A, Macarenco A V (1981) Chem. Phys. Lett. 83: 142
8. Lundin A A, Macarenco A V (1984) Sov. Phys. JETP 87: 999
9. Van Vleck J H (1948) Phys. Rev. 74: 1168
10. Bloembergen N, Rowland T J (1953) Acta Metall. 1: 731
11. Haeberlen U (1976) High resolution NMR in solids, selective averaging, Clarendon, New York
12. Maricq M M, Waugh J S (1979) J. Chem. Phys. 70: 3300
13. Hempel G, Schneider H (1982) Pure & Appl. Chem. 54: 635
14. Vander Hart D L, Earl W L, Garroway A N (1981) J. Magn. Reson. 44: 361
15. Hempel G (1980) Aussagemöglichkeiten der hochauflösenden ^{13}C-NMR für polymere Festkörper, Thesis, Merseburg
16. Spiess H W (1983) Coll. & Polymer. Sci. 261: 193; b. Spiess H W (1985) Deuteron NMR—A new tool for studying chain mobility and orientation in polymers, in: Advances in Polymer Science, Springer, Berlin, Heidelberg, New York
17. Sillescu H (1982) Pure & Appl. Chem. 54: 619
18. Andrew E R, Bradbury A, Eades R G (1958) Arch. Sci. 11: 223
19. Lowe I J (1959) Phys. Rev. Lett. 2: 285
20. Schaefer J, Stejskal E O, Buchdahl R (1977) Macromolecules 10: 384
21. Eckman R, Alla A, Pines A (1980) J. Magn. Res. 41: 440
22. Waugh J S, Huber L M, Haeberlen U (1968) Phys. Rev. Lett. 20: 180
23. Mehring M (1976) High resolution NMR spectroscopy in solids, in: NMR, Basic Principles and Progress, Springer, Berlin, Heidelberg, New York
24. Pines A, Gibby M G, Waugh J S (1973) J. Chem. Phys. 59: 569
25. Goldman M (1970) Spin temperature and nuclear magnetic resonance in solids, Clarendon, Oxford
26. Provotorov B N (1961) Soviet Physics JETP 41: 1582
27. Bloembergen N (1949) Physica 16: 386
28. Khutsishvili G R (1965) Usp. Fis. Nauk 87: 211; (1968) Usp. Fis. Nauk 96: 441
29. Aleksandrov I V, Gabrielyan R G (1977) Phys. Stat. Sol. (b) 83: K 173
30. Hunt B I, Powles J G (1966) Proc. Phys. Soc. 88: 513
31. Scheffel R (1983) Wiss. Zeitschr. Päd. Hochschule Erfurt 19: 139
32. Garroway A N, Vanderhart D L, Earl W L (1981) Phil. Trans. R. Lond. A299: 609
33. Cheung T T P, Yaris R (1980) J. Chem. Phys. 72: 3604
34. Mansfield P, Ware G (1968) Phys. Rev. 168: 318
35. Walstedt R E (1965) Phys. Rev. 138A: 1096
36. Fedotov V D, Chernov V M (1977) Vysokomol. soed. A19: 1501
37. Kadievsky G M, Fedotov V D, Grafiatullin R G (1973) Dokl. Akad. Nauk SSSR 210: 140
38. Fedotov V D, Kadievskii G M (1978) Vysokomol. soed. A20: 1565
39. Fujimoto K, Nishi T, Kado R (1972) Polymer Journ. 3: 448

40. Schenk W, Ebert A (1980) Acta Polymerica 31: 41
41. Gründer W, Schmiedel H, Freude D (1971) Ann. Phys. (Leipzig) 27: 409
42. Hiller W (1985) Theoretische und experimentelle Untersuchungen zum kernmagnetischen Relaxationsverhalten in Polymerschmelzen, Thesis, Merseburg
43. Ivanov J N, Provotorov B N, Feldman E B (1978) Soviet Physics JETP 75: 1847
44. Eroveev L N, Shumm B A, Manelis G B (1978) Soviet Physics JETP 75: 1837
45. Vega A J, Vaughan R W (1978) J. Chem. Phys. 68: 1958
46. Zobov V E, Ponomarenko A V (1981) Polym Bull. 5: 347
47. Zobov V E, Ponomarenko A V (1979) Vlijanie medlennych teplovych dvizhenii na shizniu linii JaMR v tverdych telach, in: Lundin A G (ed) Radiospektroskopija tverdovo tela, Krasnojarsk, p. 70
48. Hitrin A K (1979) Mnogoimpulsnyi spin-locking pri nalichii v obraztse medlennych molekularnych dvizhenii, in: Lundin A G (ed) Radiospectroskopija tverdovo tela Krasnojarsk, p 87
49. Hartmann S R, Hahn E L (1962) Phys. Rev. 128: 2042
50. Richter R, Hempel G, Schneider H (1978) Plaste und Kautschuk 25: 625
51. Solomon I (1955) Phys. Rev. 99: 559
52. Kuhlmann K F, Grant D M, Harris R K (1970) J. Chem. Phys. 52: 3439
53. Doddrell D, Glushko V, Allerhand A (1972) J. Chem. Phys. 56: 3683
54. Bloembergen N, Purcell E M, Pound R V (1948) Phys. Rev. 73: 649
55. Aleksandrov J V (1975) Teorija magnitnoi relaksatsii, Nauka, Moskva
56. Aleksandrov J V, Khazanovich T N (1982) in: Teoreticheskije problemy v chimicheskoi fizike, Nauka, Moskva, p 290
57. Khazanovich T N, Mironov V D (1985) Molecular Physics 55: 145
58. Chernov V M, Fedotov V D (1981) Vysokomol. soed. A23: 932
59. Fedotov V D, Chernov V M, Khazanovich T N (1978) Vysokomol. soed. A20: 919
60. Fedotov V D (1981) Impulsnyi JaMR v blochnych polymerach. Doctoral thesis, Kazan
61. Fedotov V D (1986) NMR relaxation and molecular dynamics in polymers and biopolymers: proceedings 7th European Symposium on Spectroscopy of Polymers, Teubner, Leipzig
62. Ribeiro A A, King R, Restivo C. Jardetzky O J (1980) J. Am. Chem. Soc. 102: 4040
63. Lipari G, Szabo A (1982) J. Am. Chem. Soc. 104: 4546; (1982) J. Am. Chem. Soc. 104: 4559
64. Resing H A (1967–68) Adv. Molec. Relax. Processes 1: 109
65. Miyake A (1958) J. Polym. Sci. 28: 476
66. Connor T M (1964) Trans. Far. Soc. 60: 1574
67. Blicharska B, Blicharski J S (1972) Acta Physica Polon. A41: 347
68. Grigorjev V P, Maklakov A I (1973) Vysokomolec. soed. A15: 2576
69. Grigorjev V P, Maklakov A I, Derinovskyi V S (1974) Vysokomolec. soed. B16: 737
70. McBrierty V J, Douglass D C (1980) Phys. Reports (Rev. Section of Phys. Lett.) 63: 61; (1981) J. Polym. Sci., Macromol. Rev. 16: 295; McBrierty V J (1983) Magn. Reson. Rev. 8: 165
71. Fuoss R M, Kirkwood J G (1941) J. Am. Chem. Soc. 63: 385
72. Schneider H (to be published)
73. Davidson O W, Cole R H (1951) J. Chem. Phys. 19: 1484; Cole R H, Cole K S (1941) J. Chem. Phys. 9: 341
74. Noack R (1971) Nuclear magnetic relaxation spectroscopy, in: NMR basic principles and progress, Springer, Berlin, Heidelberg, New York vol 3 p 105
75. Rouse P E (1953) J. Chem. Phys. 21: 1272
76. De Gennes P G (1971) J. Chem. Phys. 55: 572
77. Graessly W W (1982) Entangled linear, branched and network polymer systems—Molecular theories, in: Adv. in Polym. Sci., Springer, Berlin, Heidelberg, New York, vol 47 p 67
78. Kimmich R (1977) Polymer 18: 233
79. Kimmich R (1984) Polymer 25: 187
80. Cohen-Addad J P, Duepeyre R (1983) Polymer 24: 400
81. Cohen-Addad J P (1983) Polymer 24: 1128
82. Cohen-Addad J P, Guillermo A (1984) J. Polym. Sci., Polym. Phys. Ed. 22: 931
83. Cohen-Addad J P, Feio G (1984) J. Polym. Sci., Polym. Phys. Ed. 22: 957
84. Schneider H, Hempel G (to be published)

85. Haeberlen U, Hausser R, Noack F (1963) Zeitschr. Naturforsch. 18a: 689
86. Woessner D E (1962) J. Chem. Phys. 36: 1
87. Glarum S H (1960) J. Chem. Phys. 33: 639
88. Voigt G, Kimmich R (1979) Progr. Colloid and Polym. Sci. 66: 273
89. Voigt G, Kimmich R (1980) Polymer 21: 1001
90. Kimmich R, Rosskopf E, Schnur G, Spohn K-H (1985) Macromolec. 18: 810
91. Denner P, Götz H (1984) Wiss. Zeitschr. der PH Erfurt 20: 99; (1987) Wiss, Zeitschr. der PH Erfurt 23: 61
92. Skolnick J, Yaris R (1982) Macromolec. 15: 1041 (1982); Macromolec. 15: 1046; (1983) Macromolec. 16: 266
93. Valeur B, Jerry J P, Geny F, Monnerie L (1975) J. Polym. Sci., Polym. Phys. Ed. 13: 667; (1975) J. Polym. Sci., Polym. Phys. Ed. 13: 675; (1975) J. Polym. Sci., Polym. Phys. Ed. 13: 2251
94. Williams G, Watts D C (1970) Trans. Farad. Soc. 66: 80
95. Lindsey C P, Patterson G D (1980) J. Chem. Phys. 73: 3348
96. Shore J E, Zwanzig R (1975) J. Chem. Phys. 63: 5445
97. Lösche A (1957) Kerninduktion, Deutscher Verlag der Wissenschaften, Berlin
98. Farrar T C, Becker E D (1971) Pulse and fourier transform NMR, Academic, New York
99. Roth H K, Keller F, Schneider H (1984) Hochfrequenzspektroskopie in der Polymerforschung, Akademie-Verlag, Berlin
100. Powles J G, Mansfield P (1962) Phys. Lett. 2: 58
101. Mansfield P, Ware D (1966) Phys. Lett. 22: 133
102. Ostroff E D, Waugh J S (1966) Phys. Rev. Lett. 16: 1097
103. Temnikov A N, Fedotov V D (1984) Abstracts of the conference—Magnitnyi resonans v kondensirovannych sredach, Kazan, p 137
104. Jeener J, Broekaert P (1967) Phys. Rev. 157: 232
105. Stejskal E O, Schaefer J (1975) J. Magn. Reson. 18: 560
106. Torchia D A (1978) J. Magn. Reson. 30: 613
107. Trappeniers N J, Gerritsma C J, Oosting P H (1964) Physica 30: 997
108. Connor T M (1971) in: NMR Basic Principles and Progress, Springer, Berlin, Heidelberg, New York, vol 4 p 247
109. Wardell G E, McBrierty V J (1973) Proc. R. Irish Acad. 73: 63
110. McBrierty V J (1974) Polymer 15: 503
111. Clark A H, Littford P J (1980) J. Magn. Reson. 41: 42
112. a. Vjaselev I M, Fedotov V D (1984) Abstracts of the conference—Magnitnyi resonans v kondensirovannych sredach, Kazan, p 98; b. Vjaselev I M, Fedotov V D, Schneider H, to be published
113. Bergmann K, Nawotki K (1967) Kolloid-Z.-Z. Polym. 219: 132
114. Unterforstuber K, Bergmann K (1979) J. Magn. Reson. 33: 483
115. Fedotov V D, Abdrashitova N A (1977) Vysokomol. soed. A19: 2811
116. Fedotov V D, Temnikov A N (1979) Vysokomol. soed. B21: 656
117. Goldman M, Shen L (1966) Phys. Rev. 144: 32
118. Assink R A (1978) Macromolec. 11: 1233
119. Cheung T T P, Gerstein B C (1981) J. Appl. Phys. 52: 5617
120. Cheung T T P, Gerstein B C, Ryan L M, Taylor R E, Dybowski D R (1980) J. Chem. Phys. 73: 6059
121. Bergmann K, Schmiedberger H (1980) Colloid and Polym. Sci. 258: 24
122. Aue W P, Bartholdi E, Ernst R R (1976) J. Chem. Phys. 64: 2229
123. Waugh J S (1976) Proc. Natl. Acad. Sci. USA 73: 1394
124. Hester R K, Ackermann J L, Neff B L, Waugh J S (1976) Phys. Rev. Lett. 36: 1081
125. Lippmaa E, Alla M, Tuherm T (1976) Proceedings XIX-th Congress Ampere, Heidelberg, North-Holland, Amsterdam, p 113
126. Cohen-Addad J P, Duhterian J (1973) Comp. Red. Acad. Sci. Paris 277: 1315
127. Benoit H, Rabii M (1973) Chem. Phys. Lett. 21: 466
128. Cohen-Addad J P (1974) J. Chem. Phys. 60: 2440
129. Doskočilova D, Schneider B (1974) Trekoval. J., Coll. Chech. Chem. Com. 39: 2943
130. Dybowski D R, Vaughan R W (1975) Macromolec. 8: 50

131. Fedotov V D, Chernov V M (1975) Dokl. Akad. Nauk SSSR 224: 891
132. Skirda V D, Maklakov A I, Schneider H (1978) Vysokomol. soed. A20: 1412
133. Klüver W, Ruland W (1978) Progr. Coll. Polym. Sci. 64: 255
134. Fedotov V D, Ovchinnikov J K, Abdrashitova N A, Kusmin N N (1977) Vysokomol. soed. A19: 378
135. Fedotov V D, Abdrashitova N A (1980) Vysokomol. soed. A22: 624
136. Götz H, Willing T (1982) Plaste und Kautschuk 29: 661
137. Temnikov A N, Fedotov V D, Logunov V M, Finkel E E (1983) Vysokomol. soed. A25: 1086
138. Fedotov V D, Chernov V M (1979) Vysokomol. soed. B21: 656
139. Chujo R (1963) J. Phys. Soc. Japan 18: 124
140. a. Hvon S H, Horii F, Kitamaru R (1977) Bull. Inst. Chem. Res. Kyoto Univ. 55: 248; b. Kitamaru R, Horii F (1978) Adv. Polym. Sci. 26: 137; c. Horii F, Kitamaru R (1981) J. Polym. Sci., Polym. Phys. Ed. 19: 109
141. Lundin A A, Provotorov B N (1976) Sov. Phys. JETP 70: 2201
142. Lifshitz M J, Komarov E V (1983) Vysokomol. soed. A25: 2611
143. Gotlib J J, Lifshitz M J, Shevelev V A, Lishansky I C, Balanina I V (1976) Vysokomol. soed. A18: 2299
144. Chernov V M (1980) Jadernaja magnitnaja relaksatsija i molekularnoje dvizhenije v amorfnych polymerach, Thesis, Kazan
145. Gotlib J J, Lifshitz M J, Shevelev V A (1978) Vysokomol. soed. A20: 413
146. a. Geschke D, Pöschel K, Ludwigs K (1981) Polym. Bull. 5: 341; b. Geschke D, Pöschel K, Doskočilova D, Schneider B (1985) Acta Polymerica 36: 645; c. Geschke D, Pöschel K (1986) Coll. & Polym. Sci. 264: 482
147. Fedotov V D, Chernov V M, Volvson V U (1978) Vysokomol. Soed. B20: 679
148. a. Doskocilova D, Schneider B (1978) Adv. in Coll. & Interf. Sci. 9: 63; b. Doskocilova D, Schneider B (1982) Pure and Appl. Chem. 54: 575
149. Hiller W, Schenk W (1986) Polymer 27: 1353
150. Fedotov V D, Abdrashitova N A (1979) Vysokomol. soed. A21: 2275
151. Crist B, Peterlin A (1969) J. Polym. Sci., A2: 1165
152. Haeberlen U (1968) Kolloid-Z., Z Polym. 225: 15
153. Fedotov V D, Ebert A, Schneider H (1981) Phys. Stat. Sol. (a) 63: 209
154. Khazanovich T N (1969) Molecul. Phys. 17: 281
155. Woessner D E (1961) J. Chem. Phys. 35: 41
156. Douglass D C, McBrierty V J (1971) J. Chem. Phys. 54: 4085
157. Fedotov V D, Kadievsky G M, Gafiatullin R G (1974) Dokl. Akad. Nauk SSSR 217: 876
158. Kosfeld R, v Mylius U (1971) in: NMR basic principles and progress, Springer, Berlin, Heidelberg, New York, vol 4 p 181
159. Pfeifer H (1972) in: NMR basic principles and progress, Springer, Berlin, Heidelberg, New York, vol 7 p 53
160. Gerstein B C (1983) Analyt. Chem. 55: 781A
161. Gerstein B C (1983) Analyt. Chem. 55: 899A
162. Schneider H, Hempel G (1978) Proceedings XXth Congress Ampere, Tallinn, p 94
163. McConnell H M (1957) J. Chem. Phys. 27: 226
164. Havens J R, Koenig J L (1983) Appl. Spectroscopy 37: 226
165. Garroway A N, VanderHart D L, Earl W L (1981) Phil. Trans. R. London A299: 609
166. Earl W L, VanderHart D L (1979) Macromolecules 12: 762
167. Dalling D K, Grant D M (1974) J. Am. Chem. Soc. 96: 1827
168. Resing H A, Stoltfeldt-Ellington D (1980) J. Magn. Reson. 38: 401
169. Hempel G (1979) Plaste und Kautschuk 26: 361
170. Schneider H, Hempel G (1980) Wiss. Zeitschr. TH Leuna-Merseburg 22: 407
171. Spiess H W (1982) in: Ward I M (ed) Developments in oriented polymers, Appl. Science Publishers, vol 1, London
172. Spiess H W (1985) Pure & Appl. Chem. 57: 1617
173. Hentschel R, Sillescu H, Spiess H W (1981) Polymer 22: 1516
174. Opella S J, Waugh J S (1977) J. Chem. Phys. 66: 4919
175. VanderHart, D L (1976) J. Magn. Reson. 24: 467

176. Waugh J S (1976) Proc. Natl. Acad. Sci. USA 73: 1394
177. Schaefer J, Sefcik M D, Stejskal E O, McKay R A (1981) Macromolecules 14: 280
178. Hempel G, Richter R, Schneider H (1980) Plaste und Kautschuk 27: 128
179. Richter R, Schneider H (1982) Plaste und Kautschuk 29: 689
180. Richter R (1982) Aufbau und Erprobung eines ^{13}C-NMR Hochauflösungsverfahrens für feste Polymere, Thesis, Merseburg
181. DuBois Murphy P, Taki T, Gerstein B C (1982) J. Magn. Reson. 49: 99
182. Maricq M M, Waugh J S (1979) J. Chem. Phys. 70: 3300
183. Herzfeld J, Berger A E (1980) J. Chem. Phys. 73: 6021
184. Lippmaa E, Alla M, Tuherm T (1976) Proceedings XIXth Congress Ampere, Heidelberg, p 113
185. Lippmaa E (1976) Usp. Fiz. Nauk 120: 512
186. Waugh J S, Maricq M M, Cantor J (1978) J. Magn. Reson. 29: 183
187. Jelinski L W (1981) Macromolecules 14: 1341
188. Henrichs P M: cited in [164]
189. Sergot P, Laupretre F, Louis C, Virlet J (1981) Polymer 22: 1150
190. Schröter B, Posern A (1981) Makromol. Chem. 182: 675
191. Torchia D A (1978) J. Magn. Reson. 30: 613
192. VanderHart D L (1979) Macromolecules 12: 1232
193. Opella S J, Frey M H, Cross T A (1979) J. Am. Chem. Soc. 101: 5856
194. Schröter B, Posern A (1982) Makromol. Chem. Rapid Communications 3: 623
195. Schaefer J, Stejskal E O, Steger T R, Sefcik M D, McKay R A (1980) Macromolecules 13: 1121
196. Spiess H W (1985) Adv. Polym. Sci. 88: 23
197. Sillescu H, Zimmer G (1979) Ber. Bunsenges. Phys. Chem. 83: 396
198. Haeberlen U (1985) Magn. Res. Rev. 10: 81
199. Lindner P, Rössler E, Sillescu H (1981) Makromol. Chem. 182: 3653
200. Schaefer J, Stejskal E O, Buchdahl R (1977) Macromolecules 10: 384
201. Balimann G E, Groombridge C J, Harris R K, Packer K J, Say B J, Tanner S F (1981) Phil. Trans. R. Soc. Lond. A299: 643
202. Lyerla J R (1979) in: Shen M (ed) Contemp. Top. Polym. Sci., Plenum, vol 3 p 143
203. Stejskal E O, Schaefer J, Sefcik M D, Jacob G S, McKay R A (1982) Pure & Appl. Chem. 54: 461
204. Sefcik M D, Stejskal E O, McKay R A, Schaefer J (1979) Macromolecules 12: 423
205. Schaefer J, Stejskal E O (1976) J. Am. Chem. Soc. 98: 1031
206. Resing H A, Garroway A N, Weber D C, Ferraris J, Stoltfeldt-Ellington D (1982) Pure & Appl. Chem. 54: 595
207. Cholli A, Ritchey W, Koenig J L (1983) Spectroscopy Lett. 16: 51
208. Möller M, Cantow H-J, Krüger J K, Höcker H (1981) Polym. Bull. 5: 125
209. Gronski W, Hasenhindl A, Limbach H H, Möller M, Cantow H-J (1981) Polym. Bull. 6: 93
210. Havens J R, Ishida H, Koenig J L (1981) Macromolecules 14: 1327
211. Maricq M M, Waugh J S, MacDiarmid A G, Shirakawa H, Heeger A J (1978) J. Am. Chem. Soc. 100: 7729
212. McConnell H M (1958) J. Chem. Phys. 28: 430
213. Sefcik M D, Stejskal E O, McKay R A, Schaefer J (1979) Macromolecules 12: 423
214. Alemany L B, Grant D M, Pugmire R J, Alger T D, Zilm K W (1983) J. Am. Chem. Soc. 105: 2133; (1983) J. Am. Chem. Soc. 105: 2142
215. Resing H A, Garroway A N, Hazlett R N (1978) FUEL 57: 450
216. Lyerla J R, Fyfe C A, Yannoni C S (1979) J. Am. Chem. Soc. 101: 1351
217. Möller M, Cantow H-J (1981) Polym. Bull. 5: 119
218. Gronski W, Möller M, Cantow H-J (1982) Polym. Bull. 8: 503
219. Sefcik M D, Schaefer J, Stejskal E O, McKay R A (1980) Macromolecules 13: 1132
220. Jonscher A K (1975) Colloid and Polym. Sci. 253: 231
221. Saito S, Nakajima J (1958) J. Soc. Reol. (Japan) 8: 313
222. Sacher E (1968) J. Polym. Sci. 6 (A2): 1935
223. Lyerla J R, Yannoni C S (1983) J. Research and Development 27: 302
224. Wobst M (1985) Kernmagnetische Relaxationsuntersuchungen zum molekularen Bewegungsverhalten in festem, modifiziertem Polymethylmethacrylat,Thesis, Merseburg
225. Wobst M (1985) Acta Polymerica 36: 492

226. Schneider H, Wobst M, Hempel G (1984) Wiss. Zeitschr. PH Güstrow 22: 205
227. Macho V, Kendrick R, Yannoni C S (1983) J. Magn. Reson. 52: 450
228. McCrum N G, Read B E, Williams G (1967) Inelastic and dielectric effects in polymer solids, Wiley, London
229. Froix M F, Williams D J, Goedde A O (1976); Macromolecules 9: 354; (1976) Macromolecules 9: 81
230. Schaefer J, Stejskal E O, Sefcik M D, McKay R A (1981) Phil. Trans. R. Soc. London A299: 593
231. Donth E (1981) Glasübergang, Akademie-Verlag, Berlin
232. Fedotov V D, Griebel G, Schneider H, to be published
233. Donth E, Schenk W, Ebert A (1979) Acta Polymerica 30: 540
234. Schenk W, Donth E (1980) Plaste und Kautschuk 27: 12
235. Schenk W, Ebert A (1980) Acta Polymerica 31: 41
236. Schenk W, Schneider H (1982) Plaste und Kautschuk 29: 685
237. Williams M L, Landel R F, Ferry J D (1955) J. Am. Chem. Soc. 77: 3701
238. Schenk W (1980) Kernmagnetische Relaxationsuntersuchungen an PVC und am System PVC/VC, Thesis, Merseburg
239. Schneider H, Schenk W, Griebel G (1978) Wiss. Zeitschr. TH Leuna-Merseburg 20: 227
240. Gutowsky H S, Saika A, Takeda M, Woessner D S (1957) J. Chem. Phys. 27: 534
241. Slichter W P, Davis D D (1964) J. Appl. Phys. 35: 3103
242. Lenk R, Cohen-Addad J P (1970) Solid State Commun. 8: 1869
243. Connor T M (1970) J. Polym. Sci. 8 (A2): 191
244. Shevelev V A (1971) Vysokomol. soed. A13: 2316
245. McCall D W, Falcone D R (1970) Trans. Faraday Soc. 66: 262
246. Hoch M J R, Bovey F A, Davis D D, Douglass D C, Falcone D R, McCall D W, Slichter W P (1971) Macromolecules 4: 712
247. Wobst M (private communication)
248. Slichter W P (1971) in: NMR basic principles and progress, Springer, Berlin, Heidelberg, New York, vol 4 p 209
249. Hill R M, Dissado L A (1984) J. Polym. Sci., Polym. Phys. Ed. 22: 1991
250. Ferry J D (1970) Viscoelastic properties of polymers, Wiley, New York
251. Folland R, Charlesby A (1979) Europ. Polym. J. 15: 953
252. Kimmich R, Koch H (1980) Colloid & Polym. Sci. 258: 261
253. **Hiller W, Schneider H (1988) Acta Polymerica 39: 276**
254. Kimmich R (1985) Helv. Phys. Acta 58: 102
255. Dechter J J, Axelson D E, Dekmezian A, Glotin M, Mandelkern L (1982) J. Polym. Sci., Polym. Phys. Ed. 20: 641
256. a. Geschke D, Pöschel K, Eckhardt G, Brauer E (1984) Acta Polymerica 35: 269; b. Geschke D, Quillfeld E (1985) J. Magn. Res. 65: 326
257. Hempel G, Schneider H (1981) Polym. Bull. 6: 7
258. Hentschel D, Sillescu H, Spiess H W (1981) Macromolecules 14: 1605
259. Pechhold W, Blasenbrey S, Woerner S (1963) Kolloid Z. – Z. Polym. 189: 14
260. Peterlin-Neumaier R, Springer T (1976) J. Polym. Sci., Polym. Phys. Ed. 14: 1351
261. Saito N, Okano K, Iwayangi S, Hideshima T (1963) Molecular motion in solid state polymers, in: Solid state phys., Academic, vol 14 p 344
262. Zbinden R (1966) Infrakrasnaja spektroskopia polimerov, Mir, Moskva
263. Punkkinen H, Ingman L P (1978) Phys. Stat. Sol. (a) 46: 213
264. Fedotov V D, Abdrashitova N A (1985) Vysokomol. soed. A27: 263
265. Kuzmin N N, Ovchinnikov J K, Bakeev N F (1980) Vysokomol. soed. A22: 1372
266. Gotlib J J, Darinskij A A, Svetlov J E (1986) Fisicheskaja kinetika makromolekul, Chimija, Leningrad, p 29
267. Spiess H W (1978) Rotation of molecules and nuclear spin relaxation, in: NMR, basic principles and progress, Springer, Berlin, Heidelberg, New York
268. Jelinski L W (1986) Deuterium NMR of solid polymers, in: High resolution NMR spectroscopy of synthetic polymers in bulk, Verlag Chemie, Weinheim
269. Spiess H W, Sillescu H (1981) J. Magn. Res. 42: 381

270. Spiess H W (1980) J. Chem. Phys. 72: 6755
271. Lausch M, Spiess H W (1983) J. Magn. Res. 54: 466
272. Spiess H W (1983) J. Mol. Struct. 111: 119
273. Schmidt C, Kuhn K J, Spiess H W (1985) Prog. Colloid Polym. Sci. 71: 71
274. Pschorn U, Spiess H W, Hisgen B, Ringsdorf H (1986) Makromol. Chem. 187: 2711
275. Ebelhaeuser R, Spiess H W (1985) Ber. Bunsenges. Phys. Chem. 89: 1208
276. Sillescu H (1986) Makromol. Chem., Macromol. Symp. 1: 39
277. Hentschel D, Sillescu H, Spiess H W (1984) Polymer 25: 1078
278. Roessler E, Sillescu H, Spiess H W (1985) Polymer 26: 203
279. Schmidt C, Wefing S, Bluemich B, Spiess H W (1986) Chem. Phys. Lett. 130: 84
280. Boeffel C, Spiess H W, Hisgen B, Ringsdorf H, Ohm H, Kirste R G (1986) Makromol. Chem.,
     Rapid Commun. 7: 777
281. Harbison G S, Spiess H W (1986) Chem. Phys. Lett. 124: 128
282. Harbison G S, Vogt V-D, Spiess H W (1987) J. Chem. Phys. 86: 1206
283. Torchia D A, Szabo A (1982) J. Magn. Res. 49: 107
284. Sillescu H (1982) in: Benoit H, Rempp P (eds) IUPAC Macromolecules, Pergamon, Oxford
285. Cholli A L, Dumais J J, Engel A K, Jelinski L W (1984) Macromolecules 17: 2399
286. Hergeth W D, Steinau J U, Bittrich H J, Simon G, Schmutzler K, Polymer, in press
287. Schneider H, Hiller W, Fedotov V D, J. Polym. Sci, to be published
288. Feldman JuD, Fedotov V D (1987) Sov. J. Phys. Chem. 61: 2001
289. Feldman JuD, Fedotov V D (1988) Chem. Phys. Lett. 143: 309
290. Fedotov V D, Kivaeva L S (1987) J. Biomol. Struct. & Dyn. 4: 599
291. Fedotov V D (1983) Proceedings "HF-Spektroskopie an Polymeren", Reinhardsbrunn, p 38
292. Schneider H (1983) Proceedings "HF-Spektroskopie an Polymeren", Reinhardsbrunn, p 4
293. Earl W L, VanderHart D L (1981) Macromolecules 14: 570
294. Hempel G, Schneider H, Fedotov V D (1987) Wiss. Zeitschr. TH Leuna-Merseburg 29: 365
295. Steger T R, Schaefer J, Stejskal E O, McKay R A (1980) Macromolecules 13: 1127
296. Skolnick J, Perchack D, Yaris, R, Schaefer J (1984) Macromolecules 17: 2332
297. Sefcik M D, Schaefer J, Stejskal, E O, McKay R A (1980) Macromolecules 13: 1132
298. VanderHart D L (1979) Macromolecules 12: 1232
299. Kärger J, Pfeifer H, Heink W: Adv. Magn. Res. 12: (in press)
300. De Gennes P G (1979) Scaling concepts in polymer physics, Cornell University Press, Ithaka
301. Doi M, Edwards S F (1978) J. Chem Soc., Farad. Trans. 74: 1789
302. v Meerwall E D (1984) Adv. Polym. Sci. 54: 1; (1985) Rubber Chem. Technol. 58: 527
303. Maklakov A I, Skirda V D, Fatkullin N F (1986) Samodiffusia v rastvorje i rasplave polimerov,
     Kazan, State University
304. Fleischer G, Geschke D, Kärger J, Heink W (1985) J. Magn. Reson. 65: 429
305. Skirda V D, Sevrjugin V A, Maklakov A I (1983) Dokl. Akad. Nauk SSSR 269: 638
306. Sevrjugin V A, Skirda V D, Maklakov A I (1986) Polymer 27: 290
307. a. Fleischer G (1983) Polym. Bull. 9: 152; b. Fleischer G (1984) Coll. & Polym. Sci. 26: 919;
     c. Fleischer G (1985) Polymer 26: 1677
308. Bachus R, Kimmich R (1983) Polymer 24: 964
309. a. Antonietti M, Coutandin J, Sillescu H (1986) Macromolecules 19: 793; b. Green P F, Kramer E J
     (1986) Macromolecules 19: 1108
310. Fleischer G, Straube E (1985) Polymer 26: 241
311. Maklakov A I, Sevrjugin V A, Skirda V D, Fatkullin N F (1984) Vysokomol. soed 25A:
312. v Meerwall E D, Amis E J, Ferry J D (1985) Macromolecules 18: 260
313. Geschke D, Fleischer G, Straube E (in press) Polymer
314. Sundukov V I, Skirda V D, Maklakov A I (1985) Polym. Bull. 14: 153
315. McBrierty V J (1979) Faraday Discussions, Chem. Soc. 68: 78
316. Douglass D C (1980) ACS Symposium Series, No. 142, p. 147, Polymer Characterization by ESR
     and NMR
317. McBrierty V J, Douglass D C, Furnkawa T (1982) Macromolecules 15: 1063
318. McBrierty V J, Douglass D C, Kwei T K (1978) Macromolecules 11: 1265
319. Simon G, Schneider H, Häusler K G (1988) Progr. Colloid Polym. Sci. 78: 1

320. Simon G, Birnstiel A, Schimmel K H, Polymer Bulletin, to be published
321. Stejskal E O, Schaefer J, Sefcik M D, McKay R A (1981) Macromolecules 14: 275
322. Schaefer J, Sefcik M D, Stejskal E O, McKay R A (1981) Macromolecules 14: 188
323. Kwei T K, Nishi T, Roberts R F (1974) Macromolecules 7: 667
324. Nishi T, Wang T T, Kwei T K (1975) Macromolecules 8: 227
325. Douglass D C, McBrierty V J (1978) Macromolecules 11: 766
326. Stoll G (1984) Kernmagnetische Relaxationszeituntersuchungen an PMMA-PVAc-Mischungen, Thesis, Merseburg
327. Reichert D (1985) Untersuchungen zur Verträglichkeit in PMMA-PVAc-Mischungen mit Methoden der kernmagnetischen Relaxation, Thesis, Merseburg
328. Wobst M (in preparation)
329. Wobst M (1988) J. Polym. Sci. 26: 527
330. Simon G (private communication)
331. Litvinov V M, Shdanov A A (1985) Dokl. Akad. Nauk SSSR 283: 1233
332. Denner P (1988) Protonenrelaxation von Polymeren durch begrenzte translatorische Diffusion von Kettendefekten. Doctoral thesis, Erfurt
333. Griebel G (1982) Struktur und Dynamik im schlagzähen Polyvinylchlorid. Doctoral thesis, Merseburg
334. Fedotov V D, Straube E (to be published)
335. Jardetzky O (1981) Accounts of Chem. Research 14: 291
336. Jardetzky O, Roberts G C K (1981) NMR in molecular biology, Academic, New York
337. Opella S J (1982) Am. Rev. Phys. Chem. 33: 533
338. Wagner G (1983) Quart. Rev. Biophys. 16: 1
339. Fedotov V D (1983) Sov. Molec. Biol. 17: 403
340. Kivaeva L S, Fedotov V D (1984) Proceedings of the 5th conference on spectroscopy of biopolymers, Charkov, p 111
341. Fedotov V D, Kivaeva L S (1985) Proceedings of the 7th European symposium on polymer spectroscopy, Dresden
342. Fedotov V D, Kivaeva L S (1987) Sov. Biopolym. i kletki
343. Litvinov V M, Wobst M, Reichert D, Schneider H, Ždanov A A (1988) Acta Polymerica 39: 243
344. Wang C C, Pecora R (1980) J. Chem. Phys. 72: 5333
345. Howarth O W (1978) J. Chem. Soc., Faraday Trans. 74: 1031
346. Lippmaa E T (private communication)
347. Fedotov V D (1984) Proceedings of the 5th conference on spectroscopy of biopolymers, Charkov, p 248
348. Fedotov V D, Kivaeva L S, Temnikov A N, Abaturov L V, Lebedev J O (1985) Proceedings of the symposium on physico-chemical properties of biopolymers in solutions and cells, Pushchino, p 115
349. Ivannikov A I, Abramov V M, Volkov V J, Zavjalov V P (1984) Sov. Molec. Biol 17: 734
350. Fedotov V D, Feldman J D, Judin I D (1985) Studia Biophys. 107: 83 .
351. Cantor C R, Schimmel P R (1983) in: Freman W H (ed) Biophysical chemistry, San Francisco vol 3

Subject Index